Encounters with
ARCTIC ANIMALS

*Photography & text by
Fred Bruemmer*

McGraw-Hill Ryerson Limited/Toronto

ENCOUNTERS WITH ARCTIC ANIMALS

Copyright © Fred Bruemmer, 1972

All rights reserved. No part of this publication may be reproduced, stored in a retrieval system, or transmitted in any form or by any means, electronic, mechanical, photo-copying, recording, or otherwise, without the prior written permission of McGraw-Hill Ryerson Limited.

ISBN 0-7700-0344-3

Library of Congress Catalog Card Number 78-153758

1 2 3 4 5 6 7 8 9   SM-72   10 9 8 7 6 5 4 3 2

Printed and bound in Canada

Every reasonable care has been taken to trace and give credit to all owners of copyrighted material used in this book. The author and the publisher will welcome information enabling them to rectify any errors or omissions.

Grateful acknowledgment is made to Dalton Muir for the use of the photograph of gyrfalcon on page 162.

McGRAW-HILL RYERSON LIMITED
*Toronto   Montreal   New York   London   Sydney
Johannesburg   Mexico   Panama   Düsseldorf   Singapore
Rio de Janeiro   Kuala Lumpur   New Delhi*

To Aurel and René
with the hope that when they are grown up
the Arctic will still be rich in wildlife

Contents

8 Preface
10 Of Seals & Man
22 Death on the Ice
42 Islands of Seals
60 "Bestes à la grande dent"
74 Nanook - The Great White Bear
102 The Little White Fox
110 "Ill shapen beast"
126 The Living Barrens
162 "A ger-fauk that is milke white"
172 Birds of the Sun

192 The Downy Ducks
222 The Sea Unicorn
240 The Lazy Shark
246 Lake of the Cannibal Cod
254 Notes & Bibliography

Preface

By the year 2000, less than thirty years from now, there will be seven billion people on earth, according to scientific estimates. Cities will be larger and more crowded, the environment probably even more polluted than it is now and the remaining wildlife in even graver danger that at present. Already much of Asia's wildlife has been wiped out, with remnant populations surviving only in parks and sanctuaries. Of Africa's former wildlife wealth, only 1 per cent survives, again primarily in national parks. In the future, wildlife may survive only in sanctuaries, on isolated islands, in the sea, and in the vastness of Arctic and Antarctic, the two regions unsuited to agriculture.

Long ago, man in the Arctic lived in balance and in harmony with nature. This ancient balance was broken when white men came to the north to explore and to exploit its immense wildlife wealth. The great Greenland whales were nearly exterminated; of the three million caribou that once migrated across taiga and tundra, perhaps three hundred thousand are left; walrus were slaughtered for fat, ivory and hides until only a tenth of their former numbers remain.

In recent years, though, the situation has changed. Thanks to complete protection since 1917, muskoxen have increased to more than ten thousand. After decades of decline, caribou are increasing again. Most Eskimos now live in villages and hunting pressure has become localized: intense near the settlements and negligible in the now uninhabited regions.

Now a more insidious danger threatens the North and its wildlife. The discovery of oil at Prudhoe Bay on Alaska's north slope in 1968, and at Atkinson Point in the Mackenzie River delta in 1970, has redoubled industry's interest in the Arctic as a potential storehouse of infinite mineral wealth. Because of the Arctic's delicately balanced ecology, damage done during exploration and exploitation of this mineral wealth may endanger the survival of arctic wildlife.

With care and caution, and a great deal of research into the consequences of every action before the action is taken, we can have in the North both a flourishing industry and an unspoiled environment. But if the land is damaged now, and the sea polluted, the North's great wildlife wealth will vanish. The Arctic, with its exquisitely balanced nature, honed to perfection in a million-year process of selection and adaptation, is an unforgiving land. The mistakes we make there may be final – absolute and irreversible.

During the past seven years, I have spent about six months of every year in the Canadian Arctic, and previously I had made several trips to the European Arctic. On several trips I have accompanied field parties from the Arctic Biological Station of the Fisheries Research Board of Canada, and of the Canadian Wildlife Service, and the generously shared knowledge of their scientists and technicians has given me a better understanding of arctic wildlife. However, the interpretation of material and the opinions expressed in this book are entirely mine.

For their help, advice and friendship I am particularly grateful to Dr. Arthur Mansfield, Dr. David Sergeant, Mr. Brian Beck, Mr. Wybrant Hoek and Mr. Ingram Gidney of the Arctic Biological Station, Fisheries Research Board of Canada; Dr. Andrew H. Macpherson, Dr. Charles Jonkel, Mr. Gerald R. Parker, Mr. Frank L. Miller, and Mr. Charles Dauphiné of the Canadian Wildlife Service; and to Dr. Ian A. McLaren of Dalhousie University, Halifax.

F.B.

Of Seals & Man

The life and economy of the Eskimos, throughout their immense seven-thousand-mile range, from eastern Siberia to East Greenland, was based (and still is, to some extent) on seal. The Caribou Eskimos of the Barren Grounds were an exception. They could get along without seal, but preferred not to. Many groups made annual trips to the coast to catch seal, or to barter for seal products with such coastal tribes as the Aivilingmiut of northern Hudson Bay. For all the other Eskimos, life in the Arctic would have been impossible without seal, and of all the seals the most important to them were the little ringed seal and the big bearded seal.

When Dr. William E. Taylor, of the National Museum of Canada, excavated houses of the Sadlermiut Eskimos on Southampton Island, Hudson Bay, in 1954, he found that more than 70 per cent of all mammal bones discovered were of seal. The Sadlermiut, who became extinct in 1903 after contracting a disease from visiting white whalers, lived in an area rich in walrus, caribou and whale. Yet seals, apparently, were the mainstay of their existence. This must have been basically true also for the Eskimos of other arctic regions.

Ringed seal are small (a big male can weigh up to 250 pounds), attractive and, fortunately, still numerous, despite millennia of persecution. Their total number in the circumpolar arctic seas is conservatively estimated at three million, and is put as high as six million by some scientists.

They have one, frequently fatal, weakness. They are obsessively curious. Once, walking along a beach on Spitsbergen, I came upon a little ringed seal sleeping on the shore. This was rather odd, since normally ringed seal sleep out on the ice. He tried to get away, but he was very fat and I overtook him with ease as he scrabbled frantically along with his short flippers. He squirmed and wriggled and bit me quite effectively through a thick leather glove. Once I had him firmly in my arms, he relaxed. I carried him back to the base camp of the Scottish expedition of which I, somewhat inappropriately, was a member. The little fat seal had by now resigned itself to fate, lay quietly in my arms and contented itself with the odd, and overwhelmingly fishy, burp. He blinked, looking sad and lachrymose, and the fur below his eyes was matted with tears. Seals have no tear ducts from eye to nose, so the tears just splash down their faces.

A ringed seal at Koluktoo Bay, on Baffin Island. The otic (ear) opening can be seen clearly.

This little weepy fellow, after having been admired and photographed, was returned to the sea. One would have thought after this seemingly traumatic experience, he would have been only too anxious to get as far away from us as possible. But no. He was still curious. He swam out some thirty yards, bobbed up, stared at us with his large, rather short-sighted eyes, dived and popped up a little closer, just desperately curious to find out what we were.

The Eskimos, of course, take good advantage of the seal's curiosity. They lie quietly in a boat and make scratching noises on its bottom boards. Consumed with curiosity, an incautious seal may swim within rifle range. One spring I spent long frigid days with Pewatook, an Eskimo living on Jens Munk Island in northern Foxe Basin, hunting seal at the floe edge, the limit of the landfast ice. With his harpoon shaft, Pewatook scratched across the ice, back and forth, hour after hour, with the infinite patience of the true hunter. The grating noise travels far in the water, and from time to time an inquisitive dark head bobbed up, stared fascinated, disappeared, bobbed up closer, until it strayed into the fatal range of the gun.

But even the curious and trusting little ringed seal is capable of learning caution. In isolated Koluktoo Bay, northern Baffin Island, where I spent a summer more than one hundred miles from the nearest settlement, the seals were downright silly. When we crossed the bay to check narwhal nets, curious round heads popped up all over to have a better look at the passing boat. Sometimes seals swam right into our path, surfacing a few feet from the canoe, oblivious to motor noise and danger. Later, at Ward Inlet off Frobisher Bay, an area where seals have been intensively hunted for a long time, I lived some weeks at an Eskimo camp. Here the seals behaved with utmost caution. They dived long before a boat was in shooting range, and no strange noise would lure them nearer. They knew boats and feared them, and the Eskimos sometimes spent a whole day in their boats, unable to approach a single seal.

While other arctic seals migrate south with the onset of winter, ringed seals and bearded seals remain in the north. As soon as ice begins to glaze bays and coves and inlets, and the off-shore areas the ringed seals prefer, they cut a considerable number of breathing holes into the ice cover. As the cold increases, and the ice grows thicker, the seals keep nibbling away until they may have cone-shaped shafts through ice six feet thick, vital vents to the air above. Snow soon covers these *agloos,* as the Eskimos call the breathing holes, and they become extremely difficult to find. Each seal has many breathing holes and one set may overlap that of a neighbor, thus enlarging its range even further. Its hearing is acute, and the crunching step of a man on the harsh, brittle arctic snow gives it plenty of advance warning to stay away from certain holes.

Once an Eskimo has found an agloo, sometimes with the help of dogs, he builds a windbreak of snow blocks and settles down for an incredibly

A ringed seal pup on the spring ice.

patient wait. Since the seal comes up noiselessly, the hunter may insert a thin sliver of wood or bone into the hole. In other areas Eskimos place a feather or a bit of swan's down over the hole. When the seal comes up to breathe, the stick jiggles or the down moves, and the Eskimo, taut and poised, drives his harpoon in with deadly force. Often, though, he waits in vain for hours, perhaps for an entire day. The polar bear, who lives primarily on ringed seal, is also a patient hunter. He usually enlarges the agloo sufficiently to get his paw in. Then, crouched down, he waits. If a seal surfaces, the great clawed paw strikes with lightning speed, crushing the thin-boned skull of the luckless seal. The bear grabs the carcass with his teeth and yanks it out of the constricted hole with such force that it may break all the ribs in the seal's body.

The ringed seal lives partly on small fish, such as the little polar cod, extraordinarily numerous in arctic seas; but mainly on two tiny, shrimp-like, lovely-named crustaceans, *Mysis* and *Themisto*.

At the end of March or early in April, each pregnant female ringed seal cuts a hole through the fast ice, large enough to haul out. It is usually underneath a deep snow drift, often in the lee of frozen-in icebergs or high pressure ridges. There she excavates an oval chamber, called *nunarjak* by the Eskimos, a foot or two high, about three feet wide, and sometimes twenty feet long, as a nursery for her pup. The pups weigh about ten pounds at birth. They are adorable little creatures, clothed in white, down-soft, wooly fur, with dark muzzle and great, dark, shining eyes. Until they moult their foetal "whitecoat," the pups cannot enter water. Their thick, soft coat would become waterlogged and they might drown. They lie quietly in their snow lair, now covered with a firm coating of ice created by the warmth and moisture of the immured animals.

The seal mothers nurse their pups frequently. Their milk contains at least 40 per cent fat and 10 per cent protein (as compared to 3.4 per cent fat and 3.3 per cent protein in cow's milk) and on this rich diet the pups quadruple their weight in a month. By then their moult is nearly complete. They now have a new short-haired coat, a soft-glowing silver on belly and sides, and dark grey on the back, ringed with elegant silvery rosettes. After six weeks, the young seal is ready to go to sea. Unlike other seals who lose interest in their offspring after approximately two weeks of intense feeding and solicitous maternal care, ringed seal mothers may keep their young with them and nurse them another month or more.

This is also the mating time of ringed seals, and the males in rut exude a powerful and, to humans, rather revolting smell. *Tiggaks*, the Eskimos call them – "stinkers."

Not all pups get a chance to moult and reach the safety of the sea. In March female polar bears emerge from the wintering dens with their cubs. Male polar bears and barren females are lean and hungry after months of difficult hunting in wintry darkness and cold. Now, ringed seal pups are easy prey if they can discover them in their lairs. The polar bears patrol the likeliest areas assiduously, and their keen sense of smell can detect a pup through many feet of compacted snow. The great white arctic wolves dig out the odd seal pup, and even the small arctic fox may occasionally burrow its way into a seal's lair and kill a new-born pup with its sharp fangs.

Eskimos, too, are in search of nunarjaks. They use dogs who sniff out the hidden lairs, or the Eskimos probe likely snowbanks with a sharpened stick or their harpoon shaft. If it breaks through into the hollowed-out seal nursery, they dig out the little pup. In the past, the soft fur of pups was made into warm children's clothing, sewn into underwear (worn fur side inside), or used to line mittens or boots. Now most Eskimos wear store-bought clothes and rarely spend much time searching for nunarjaks.

In April and May the seals enlarge their breathing holes and haul out to snooze in the spring sun, dark spindle-shapes on an infinite expanse of white. At this time, when the seals moult, they tend to be sleepy and lethargic. Usually solitary in their habits, they now seem to seek company. On Jones Sound I have seen as many as seventeen seals lie in a circle around one large breathing hole.

There is, no doubt, some safety in numbers. A single seal is fairly easy to approach. It sleeps for a while, awakes, looks around carefully, scratches itself languidly with its flippers, and goes back to sleep if all seems safe. Bears and Eskimos synchronize their stealthy stalk towards a seal, with its

Beautiful and guileless, a little ringed seal on an arctic beach.

(Left) A bearded seal rests on the ice. (Right) A bearded seal surfacing in Hudson Bay. Because of its broad, paddle-shaped front flippers it is often called "square-flipper."

sleep-wake periods. When the seal sleeps they noiselessly advance. The instant it moves, they freeze into immobility. The polar bear is camouflaged by its yellow-whitish fur. The Eskimo hides behind a white, shield-like hunting screen. In a group of seals, though, there is nearly always one awake. The instant it sees something move, it dives and all the other seals follow.

There can also be a disadvantage to clustering in groups around one breathing hole. Once, travelling by dog-team on Lancaster Sound, we suddenly came upon a group of seals. Our dogs, excited by the sight of something dark, moving and presumably edible, rushed forward, and the six alarmed seals dived in unison with the result that three got stuck in the hole. Hind-flippers waggled frantically in the air as the seals tried to wriggle downwards only to get more firmly wedged, and we were nearly upon them when one slipped in and the other two followed.

The great bearded seal (adult males and females can weigh up to eight hundred pounds) is even more a loner than is the ringed seal. As its name implies, it sports a set of resplendent whiskers. When they are wet they are straight and pointed. When the whiskers dry they curl at the tips,

sometimes into tight whorls like the tendrils of clinging plants. The whiskers are highly sensitive and probably help the seal to seek its food. It catches some fish, especially the small polar cod, but most of its food is gathered from the sea bottom: worms, crabs, shrimps or sea cucumbers. Its favorite food is cockles and whelks. Only the bodies are eaten, the shells the seals manage to discard. Just how they do it, no one really knows. One suggestion is that they may crush them between their flippers.

Like the ringed seal, the bearded seal inhabits the circumpolar arctic seas. Members of Fridtjof Nansen's *Fram* expedition, drifting across the Central Polar Basin, shot a bearded seal at 85 degrees north, and Russian scientists have reported it at 88 degrees north. Ringed seal, too, occur in the Central Polar Basin, and have been seen nearly to the North Pole, but neither species is common in this region. The bearded seal really likes fairly shallow water, although it can dive to considerable depth. It can, if forced to, cut breathing holes through ice and keep them open all winter; but it rather avoids this chore by staying away from fast ice altogether and seems happiest floating about on some floe, diving for food through adjoining leads whenever hungry.

The bearded seal is usually shy and not nearly as curious as the ringed seal. When a boat comes near, it rolls forwards, exposing part of the back and often dives with a splash. In summer when they moult, bearded seals

(Above) This is what one normally sees of ringed seal: an inquisitive little head. (Below) Swift and graceful in the water, ringed seal crawl only with difficulty on land.

lie on ice floes and are then rather loath to enter water. Tooma, a Southampton Island Eskimo, took me along on a bearded seal hunt in Hudson Bay. We zigzagged in his Peterhead boat through the broad leads of a loose ice pack. The seals saw us from a distance (bearded seals are not as myopic as ringed seals) but were torn between fear and their reluctance to enter water. While they dawdled in indecision death came closer. By the time a poor seal had made up its mind that the situation demanded instant evasive action it was often too late. The Eskimos' rifles blasted away, and the great seal sagged down in a fast-spreading pool of blood.

For the Eskimos, ringed and bearded seal were (and in some areas still are) the basis of their existence. The meat, boiled or raw, was the staple food of the people and their dogs. The vitamin-rich liver of ringed seals was prized as a delicacy, but the liver of the bearded seal, like that of the polar bear, contains so much vitamin A as to be toxic. If it is eaten, it can produce occasionally fatal attacks of hypervitaminosis.

The shafts of Eskimo *kamiks* (seal skin boots) are made from ringed seal skin, and the soles of strong, durable bearded seal leather. Much of the summer clothing was made of seal skins, though for winter wear the much warmer fur of caribou was preferred, whenever it could be procured.

Five bearded seal skins, scraped clean, covered a kayak, and about twenty were needed to sew the covering for the large open travelling boat, the *oomiak;* though in some areas split walrus hide was preferred.

The rich seal blubber gave the Eskimo strength, health, endurance and warmth. Rendered into oil, it burned in his crescent-shaped lamps.

The strong leather of the bearded seal was cut into thong, to serve as dog-team traces, lashings, tumplines, or to be braided into the long, tapered whip of the dog-team driver.

And when the white man came, with his insatiable desire for furs, the skins of seals became an important trading item. Many bearded seal skins, even today, are used for thong and leather, and some of the ringed seal skins are used to make boots, gloves and sledging pants. In addition, Canadian Eskimos trade in seventy thousand to a hundred thousand ringed seal pelts annually and earn, depending upon fluctuating fur prices, anywhere from a quarter of a million to over a million dollars.

In the past, Eskimos lived widely scattered in small groups along the Arctic littoral. As traders and missionaries came to the north, Eskimos began to concentrate near the newly established posts. Now nearly all of them have moved to settlements, and hunting pressure has become localized: high near the villages, negligible elsewhere. As a result, seal populations in the Arctic have decreased in some areas, such as Frobisher Bay and parts of Cumberland Sound on Baffin Island, while farther from the settlements seals hold their own or may even be increasing since the

A ringed seal near the coast of Baffin Island.

polar bear, their main enemy apart from man, has decreased.

As long as Eskimos hunted seal with kayak and harpoon, they either killed a seal or it escaped unscathed. Now hunting is done with rifles from motor-driven canoes, and the number of seal killed but lost is often alarmingly high. The bearded seal sinks as soon as it is killed. Unless it is harpooned first and then shot, or shot and immediately afterwards harpooned, the dead seal disappears before the hunter is close enough to throw the harpoon.

In spring and summer, ringed seals sink, partly because their blubber blanket at this time is thin, partly because the sea is less saline due to the influx of fresh melt water. During this time, the number of seals killed and lost is very high — from what I have seen, about a seal lost for each one taken. By the middle of August most ringed seal are sufficiently buoyant to float after they have been killed, and from then on throughout winter few seals are lost because of sinking.

The older Eskimos, many of whom can still remember the time when a lost seal might mean gnawing hunger, hunt with deadly efficiency. Akpaleeapik and Akeeagok, two brothers from Grise Fiord on Ellesmere Island, with whom I made a six-week dog-team trip, shot about thirty-five seals during that time to feed us and our dogs. They used hunting screens, stalked the seals with caution and care, and hardly ever missed. They killed so that they might eat.

The younger men, especially those living in larger settlements, have lost many of the skills, and much of the motivation, that made their fathers first-rate hunters. They shoot at distances that are too great, kill sometimes purely for the sake of killing, caring little whether a seal sinks or not, because its meat and blubber and skin are no longer vital to them, as they once were to their elders.

Both bearded seal and ringed seal are still numerous. In some areas, no doubt, their numbers have decreased. But since they are scattered over the immensity of all the arctic seas, they have escaped the fate of the more gregarious seals: to be slaughtered wholesale by the white man and his efficient sealing fleets.

For thousands of years these two seals have sustained man in the Arctic. Caribou were important to the Eskimo. In some areas, like Bering Strait, whales and walrus provided him with the bulk of his food. The Thule culture Eskimos throughout the Arctic must have relied heavily on the great Greenland whale they had learnt to hunt and kill. There are indications that long ago a group of Eskimos living in northern Ellesmere Island specialized in hunting muskoxen. All these animals contributed to maintain man throughout the vastness of the Arctic. But the basis of Eskimo life was the seal, the big bearded seal and the little ringed seal. Without them most of the Arctic would have remained a land without people.

Death on the Ice

In 1909 Dr. Wilfred T. Grenfell, the famous medical missionary to Labrador, wrote: "It has not been easy to convey to the Eskimo mind the meaning of the Oriental similies of the Bible. Thus the Lamb of God had to be translated *kotik* or young seal. This animal, with its perfect whiteness as it lies in its cradle of ice, its gentle, helpless nature, and its pathetic innocent eyes, is probably as apt a substitute, however, as nature offers."[1]

This reflection has not deterred man from slaughtering either lambs or seal pups.

No one can say with any certainty how many harp seal lived in the northern seas, before commercial exploitation of the herds began, more than two hundred years ago. Estimates range as high as thirty-five million. If this is approximately true, then the harp seal population of the world has shrunk to less than 10 per cent of its former numbers.

Harp seal are grouped into four herds, named after their breeding areas. Most seals of the "White Sea" herd spend summer and fall north of the 75th parallel, dispersed over an area from Spitsbergen in the west, to the northeastern parts of Russia's Kara Sea. In early winter, the seals migrate south to the White Sea, where the pups are born on the ice at the end of February and in early March.

Sealing in this area began in the nineteenth century, but catches at first were small. They rose sharply early in this century and reached their peak in 1925, when Norwegians and Russians killed more than half a million seals. The White Sea harp seal herd, estimated at 3,500,000 in the early 1920s, dwindled to about 1,300,000 by 1952. In 1955, a kill quota of 100,000 seals per year was established, but it was too big and came too late. In 1962 only 750,000 seals were left. Two years later Russians and Norwegians reached an agreement to suspend sealing for five years, to give the seals a chance to increase again.

Seals of the "West Ice" herd pup on the ice east of Greenland and north of the small Norwegian-owned island of Jan Mayen. Once numbered in millions, they are now reduced to about half a million. The exploitation of this herd began early in the eighteenth century as a sideline to arctic whaling. At the end of the nineteenth century, as many as 200,000 harp

seals were killed on the West Ice annually. In the years between 1950 and 1960, the average annual kill by Norwegian and Russian sealers had decreased to about 38,000.

The world's largest seal herds lived – and died – off Canada's coast. The seals of these herds spend the summer along Greenland's west coast. Some migrate as far north as Kane Basin, two thousand miles from their breeding area. Others scatter along the coast to Baffin Island into the straits separating the islands of Canada's arctic archipelago and into Hudson Bay. Harp seal are gregarious animals and much more pelagic than the coast-loving bearded seal and ringed seal. One sees them in groups of ten or twenty, arching out of the water when they surface and disappearing with a flourish of high-held hind flippers as they dive again to feed. Adult harp seals prefer fish. Young animals gorge themselves on small crustaceans.

By September the seals begin their long southward migration. In late October they pass Cape Chidley, the extreme north tip of Labrador, and by the middle of December the seals have reached the Strait of Belle Isle. Here the herds divide. About a third of all the harp seals swim through the strait into the Gulf of St. Lawrence, where they will spend the winter and where the females will haul out in early March to bear their pups on the ice. This is the "Gulf" herd. The remaining two-thirds of the seals pass the winter off the coasts of Newfoundland and Labrador, usually near the pack ice. This group is known as the "Front" herd. In late February the seals swim northward. The females seek out the best concentrations of pack ice, and on this immense mass of moving floes they give birth to their pups.

Off-shore sealing began in the seventeenth century. Early French-Canadian records are full of references to the hunt of the *loup-marin,* the "sea-wolf," as the seal was called. But settlers were few and seals numerous and, like the Eskimos in Greenland who also hunted harp seal avidly, the toll they exacted could be easily borne by the immense herds.

But early in the nineteenth century, the seal "fishery" became, next to real fishing, Newfoundland's most important industry. In 1805, 81,000 seals were killed. Eleven years later the kill had soared to 281,000 and in 1831, the all-time peak year, the Newfoundland fleet took 687,000 seals! In the years between 1830 and 1850 Newfoundland sealers killed an average of 440,000 seals each year; a total of 8,800,000 seals!

From then until World War I, the average annual Newfoundland kill was about 300,000 seals per year. In the waning days of arctic whaling (after two hundred years of ruthless pursuit, the whalers had nearly exterminated the great Greenland whale), the whalers too awoke to the fact that there was profit in seals. In 1889 whaling ships killed 303,000 seals of the Front herd, before continuing north to pursue the decimated whales. Despite the

(Left) A female harp seal hauls out on the ice to join her pup. (Below) Head raised, trilling a loud warning, the mother adopts the typical threat posture of the harp seal.

slaughter of millions of seal (maybe as many as twenty million during the nineteenth century alone), the seals of the Front and Gulf herds were still estimated at ten million in 1900.

Sealing was a great industry. In 1855, four hundred ships with thirteen thousand men sailed from Newfoundland to the breeding pack. In 1863 the first steamer bored its way into the ice and, in 1888, forty-four steamers took part in the great annual seal hunt.

Sealing was also dangerous. Caught in the shifting ice, dozens of vessels were crushed and men were often marooned on ice floes. It has been estimated that twenty thousand men from Newfoundland died in the pursuit of seals.

By the early 1930s, the seals had dwindled to about four million and to about three million by 1939. Sealing ceased during the war years and the herds began to increase again. After this brief respite the slaughter of seals was resumed with redoubled energy and made even more efficient by the

Sleepy and full after nursing, a harp seal pup snoozes on the ice.

aids of modern technology. Spotter planes located the seal herds. Helicopters ferried hunters to the floes and carried back the pelts of pups. Even Russian icebreakers forced their way into the pack. About three hundred thousand seals were killed annually in the 1950s tapering off to two hundred thousand a year in the 1960s.

The effects of this ruthless persecution soon made themselves felt. In 1951 the harp seals of the Gulf and Front herds were estimated at three and a half million. Ten years later, less than half that number were left and they have been further reduced since then. For the Gulf herd a quota of fifty thousand seal pups was established in 1965. In addition, local residents can kill seals, and they may take another thirty thousand per year. In October 1969, the Canadian government banned the killing of whitecoats in the Gulf of St. Lawrence. In future only pups that are beginning to moult, the so-called "raggedy-jackets" and fully moulted pups called "beaters" may be killed, either with a regulation-sized club or with a gun. The quota was not changed and in the spring of 1970, sealers in the Gulf quickly killed fifty thousand young seals, but the landsmen's kill was small because of the adverse weather and ice situation.

But on the Front the seals breed in international waters, and in "good" years, such as 1966, 80 per cent, and even more, of all pups have been killed. In addition, adult seals are shot in large numbers. There is a sealing season agreed to by the main participants in the hunt, Norway and Canada; and for the first time in 1971 a quota limited the kill. New agreements between Canada and Norway are likely to reduce quotas and establish maximum kill for the herds at the desirable sustained-yield level.

In the past, sealing was regarded as a romantic (which it isn't) and dangerous (which it is) occupation. Sealers were referred to as "Vikings of the Ice," and their ships were the "Wooden Walls Among the Ice." It was all rather remote and abstractly fascinating. A few people wrote books, and many read them, about the sealers' derring-do; no one really gave a hoot about the seals, except some biologists whose warnings that the herds were being dangerously decimated were larged ignored.

This pleasant picture was abruptly shattered when some Montreal cameramen filmed the famous "seal hunt" in the "Gulf" in 1964, and presented Canadian television viewers with its brutal reality.

A harp seal pup is one of the world's loveliest creatures. Its natal fur is white and woolly and silky-soft. The eyes are large and dark and lustrous. It has a black muzzle and stiff, droopy whiskers. It is innocence incarnate. In its first days, it usualy shows no fear. If a man approaches, the little seal looks trustingly at him, may even waddle towards him. It is completely captivating.

The television viewers then saw the sealers fan out over the ice, armed with gaffs and clubs, smash the pups' skulls, slit them open and rip the

Coated in silky-soft white fur, harp seal pups look at human intruders in wide-eyed innocence. They show no fear of man.

sculp, the skin plus blubber, off the twitching corpses. Worse, the film showed how sealers in their frantic haste sometimes barely stunned a pup and then skinned the squirming, screaming animal alive.

The horror and revulsion this "show" produced, grew into a world-wide campaign to "humanize" the hunt. The sealers were suddenly no longer daring "Vikings of the Ice," but branded as "butchers," "murderers" and "sadists."

Most hunting is cruel. The Eskimo who drives a harpoon into a seal and then pulls the unfortunate animal close, with the jagged barb deeply imbedded in its flesh, is cruel. The polar bear is cruel. He kills adult seals instantly because otherwise they might escape. But let a polar bear find a patch of helpless harp seal pups and he will take a bite out of one, toss another into the air, and wander through the herd leaving a trail of horribly maimed pups. Some sealers, no doubt, are needlessly cruel. They work in freezing temperature, in cutting wind, on slippery ice. They work against time. The more pups they sculp, the more money they make. They club and skin and run, and club and skin and run.

A seal's head is a small target. Often a frightened pup pulls in its head, exposing a thick, blubber-protected neck, and the blow with gaff (now outlawed) or club may merely stun it and then the animal is skinned alive. Most of the cruelty is not intentional. It is the nearly inevitable result of men working under difficult conditions, working in feverish haste to get the maximum share of the kill before the quota is filled. Government inspectors now supervise the hunt in the Gulf. But they cannot be everywhere. Unless the whole method and rhythm of the seal hunt is changed, cruelty will be part of it.

In all the furore about cruelty, another crucial aspect, that of preserving these seals, has been nearly forgotten. The Gulf herd may now hold its own. But unless protective international measures are taken (and it now seems certain that, beginning in 1972, a realistic, sustained-yield quota will limit the hunt) the Front herd will be exploited to the point where its few survivors no longer make it worth while to launch expensive sealing expeditions.

Though man may not, nature herself, it seems, is trying to help its harried seals. Studies made by Dr. David Sergeant of the Fisheries Research Board of Canada, the country's foremost expert on harp seal, show that the age of sexual maturity of female harp seal has been sharply lowered in recent years. Until about fifteen years ago, female harp seals on the Front reached reproductive age at five and a half years. Now the average age at which they breed is four years. They can thus produce an extra crop of pups. Similarly, the age of sexual maturity in female harp seals of the Gulf herd, who are subject to less intensive hunting pressure, has declined from six to five years. Harp seal may live to thirty-five years of age, and females probably bear pups until they are twenty years old. Thus, if rules for the conservation and management of the Front herd are introduced while it still contains many hundreds of thousands of seals, the herds may recuperate fairly quickly. If the Front herd receives no protection, then this seal herd, once the largest in the world, seems doomed to oblivion.

It has been suggested by conservationists that the seal herds, like the herds of animals in African parks and sanctuaries, be made into a tourist attraction. This would obviously put a stop to sealing, yet might bring in a comparable amount of revenue. If it were practicable, those visitors who might make the "seal tour" would see a spectacle they would never forget, as I shall never forget my trip to the Front herd, together with a team of five scientists and technicians from the Fisheries Research Board.

Our task was to tag as large and widespread a sample as possible of harp seal pups on the Front ice, so the survival rate of young seals might be more accurately determined. The icebreaker *d'Iberville,* on which we travelled, reached the Front ice off the southern Labrador coast on March 9.

A storm raged at sea and the vast fields of pack ice rose and fell in the heavy swell. The floes crashed together with a grim grinding noise, then drifted apart, suddenly separated by gaping leads of open water. This was the icy, rocking cradle of the new-born seals.

Using the helicopter was obviously out, but, rather than waste a day, Dr. Arthur Mansfield, leader of our party, decided to work in the area of the ship. We split into three groups. Technician Bill Shields and I went together, carrying a long aluminium ladder along to straddle the gaps. Bill held one end in place while I balanced across. Then I held the see-sawing ladder and Bill tiptoed across the icy water to the next floe.

Most seal mothers left their pups the moment we approached and the abandoned waifs immediately set up a chorus of mournful bleating. But some females, especially the younger ones, defended their pups resolutely. If we approached one of them, it raised head and neck high in its "threat posture," warning us, at the same time, in a high-pitched, trill-like, prolonged snarl to stay away. When we came too close to an aggressive female, she suddenly charged, slithering and humping over the ice with amazing speed. The seal mothers also attacked, with the same high-pitched snarl, any other female coming too close to their pups.

The pups, only a day or two old, cried plaintively when forsaken by their mothers. Although whitecoats can swim, their soft, woolly fur becomes waterlogged and they are loathe to enter water until after the moult is completed at the age of two to three weeks. As we approached, they looked at us with large limpid eyes, more curious than afraid, but cried in panic when we touched them to affix the tag to a skin fold between tail and hind flipper. Their mothers surfaced frequently in nearby leads, treading water, and watched us with bulging, worried eyes. As soon as we

Threat posture of a harp seal mother.

left, they hauled out, sniffed their softly whimpering pups, and stretched out to nurse them.

The older pups seemed to realize we represented danger. They snarled in rather pathetic miniature defiance, bit our boots and pants with sharp little teeth, and scratched us thoroughly with flailing sharp-clawed flippers.

Some pups, though, would go into a state of shock. Like frightened caterpillars, they seemed to shrink into themselves, getting shorter and chubbier, and lay on the ice in a quiet trance. Such a pup we could pet, roll over, tag, or pick up, without the least resistance, and some "entranced" pups whom I watched from a distance, would continue in this state for as long as ten minutes after we left them, although their mothers' appearance instantly snapped them out of their trance.

A few females fell into the same trance, when we approached. But an adult harp seal, well equipped with long, sharp yellowish teeth, and weighing up to four hundred pounds, is not to be trifled with, and none of us had the courage to find out how deep their trance-like sleep really is. None, that is, except Dr. Mansfield. True scientist that he is, he propounded the theory that adult seals, too, will be held in immobility by their trance, and promptly proceeded to prove this. He petted an adult seal, rolled it and tagged it, without eliciting the least response, and later repeated the feat with five other mesmerized harp seals.

The second day, our ship's helicopter discovered the main breeding patch. Two days were left until the start of the sealing season. Already more than a dozen sealing ships, mainly from Norway, were cautiously squeezing their way through the ice towards their prey. The ice had compacted overnight. Where open leads had been our hazard the previous day, crazily jumbled ice blocks, piled high into pressure ridges, now made the going difficult. The helicopter ferried us out to the main breeding area. Below, on the ice, were the seals, myriads of dark dashes on an immense white sheet. "About one hundred thousand," Dr. Mansfield said.

This particular part of the pack was at least forty square miles in extent. We were supposed to stay in sight of each other. Just in case. But with all those pressure ridges it was impossible. A couple of female seals fled and dived into a hole full of brash ice. One of the pups crawled into a crack between two ice blocks. A female surfaced, did not see me behind a sheltering ridge, hauled out and approached the nearest pup. She sniffed it and the pup responded eagerly, but was rebuffed with a cuff. At that moment its real mother appeared and, with an angry snarl attacked the female near the pup. The other pup whimpered, its mother slithered over, and peace reigned again on this little section of the ice.

When I looked up, I was alone. I wandered all day over this fantastic floating landscape of weirdly contorted ice, watching the seals lie languidly near the edge of the floes, feeding their greedy pups. The milk is rich, and the pups, who weigh twenty to twenty-five pounds at birth, balloon out to ninety pounds and more in two weeks. Unless they are frightened, the females do not leave their pups during the first two weeks, and the pups spend an idyllic life, feeding, or sleeping near their mothers, short flippers pressed to bulging bellies.

Towards evening, sudden vicious snow squalls raced across the ice, cutting visibility to a few feet. I felt lost and forlorn. The snow obscured all in a greyish moving pall, and there was nothing except the whistling of the wind in the ice blocks and the mournful mewing of the baby seals.

It was getting dark when the helicopter suddenly appeared, like a clattering ghost out of the gloom, swooped down to pick me up and homed unerringly through the blinding storm to the ship.

The last day was perfect. Sunny and calm. The pack ice, so gloomy and menacing the day before, now sparkled and shone in serene beauty. The air was clear and bright, and as far as the eye could see the white immensity of the pack was dotted with the dark spindle-shapes of seals.

In the afternoon I came across a hooded seal pup. It was nearly twice as large as the harp seal pups and its coat was a glistening blue-grey on top, and silvery white on the underside. Just as I was photographing it, I heard a snort behind me and there, in a lead, was the dark, ram's head of its mother. She dived, came up with a rush, and I took off in a hurry. Hooded seals, both male and female, are notorious for their speed and aggressiveness.

In the distance, sixteen sealing ships were ringing the breeding patch, their motors turning slowly to prevent the propellers from getting stuck in the ice, waiting for this day to end – and then the sealing season would begin. At midnight, precisely, their great searchlights flared on and the sealers scrambled over the sides of their ships. The slaughter on the Front ice had begun.

I flew over the ice for a last time next day. The sealers had worked fast. The seals had disappeared. The immense white pack was now streaked with trails of blood, where sculps had been hauled into piles. Thin little corpses dotted the ice. (At the age of two weeks the pups are so fat, the sculp – skin plus blubber – constitutes two thirds of their weight.) Of the cuddly little pups nothing remained but heaps of blood-smeared pelts. Some day, tanned and treated and dyed, they would be worked into elegant coats. The helicopter turned and flew back to the ship. I watched the blood-soaked ice below, and thought of Mencken's crusty crack, that "if God had meant women to wear fur, they would be born with it."

The great hooded seal of the arctic seas is a remote relative of the elephant seals. Like them, the male "hood" has an inflatable proboscis, a sort of flexible, bladder-like appendage, drooping from the forehead right down

Harp seals begin to moult the downy white natal fur when they are about two weeks old.

(Above) A hooded seal mother nurses her pup. Hooded seal parents are very protective and this mother, fearing her pup threatened, attacks furiously.

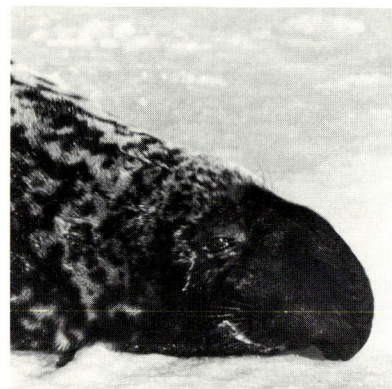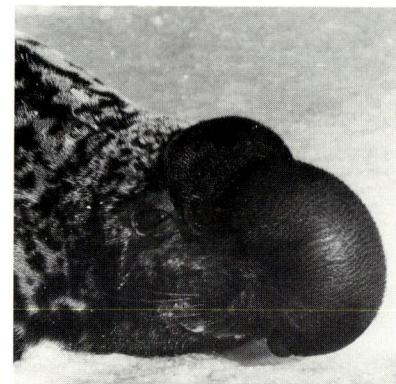

(Top, left to right) A male hooded seal dozes on the ice, its hood deflated. When excited or annoyed, it blows up its skin crest. It can also extrude its nasal septum and blow it into a large balloon. (Below) A huge bull with nasal septum blown up, and its smaller female.

over the seal's upper lip. When the seal is annoyed, and sometimes even when it is not, it blows up its proboscis into an oval crest, twice the size of a football. As if not satisfied with this startling performance, nature has equipped the male hood with an extremely large and flexible nasal septum. This he can extrude from one of the nostrils and blow up into a big, usually bright red, balloon.

Adult males reach a length of over ten feet and weigh up to eight hundred pounds. Females are about a foot shorter. Their bluish-grey coat is spattered with dark, irregular spots, and the face, beginning just above the eyes, is solid black.

Unlike the gregarious harp seal, who seems to like company at all times of the year, the hooded seal is of a more solitary disposition. But in March, the seals congregate in favorite breeding areas. Most of them seek out the "West Ice," north of Jan Mayen. Smaller groups bear their pups in roughly the same areas as the Gulf and Front herds of harp seal, on the pack ice in the Gulf of St. Lawrence and off the Labrador coast. Due to climatic changes, more than because of overhunting, these two groups have dwindled in recent years and it is believed hooded seals are deserting their breeding grounds off the Canadian coast in favor of the West Ice.

While female harp seal lie along the edge of the floes in fairly dense clusters, sometimes more than six thousand or seven thousand per square mile, hooded seals are more widely dispersed over the pack, usually in family groups, a male, a female and her pup.

The pup is only nursed for ten days to two weeks. The drifting pack is a precarious nursery. Storms may scatter the ice, or break it and crush the pups, and waves may sweep the pups off the floes. The shorter their nursing time, the smaller the danger of such disasters. But if maternal care is short, it is intense.

Both the male and the female hooded seal furiously defend their new-born pup (which doesn't bother sealers — they simply shoot them and then club the pup). The mother's milk is extremely rich and the little seal expands nearly visibly, gaining several pounds each day.

After the breeding season, the hooded seals from the Gulf and the Front swim northeastward. A few travel up Davis Strait and into Baffin Bay, as far north as the Thule district of Greenland. Most of them spend early summer on the West Greenland coast, then swim south around Cape Farewell and into Denmark Strait, between Greenland and Iceland, where they haul out on the ice to moult. To this moulting ground come also the hooded seals of the West Ice herd.

These are immense and fairly complex migrations and occasionally a hooded seal, usually a young animal, strays from the ancient path and turns up in strange locations, like Florida, for example. On the east coast of Hudson Bay, a few years ago, a Povungnituk Eskimo hunter was surprised to find a seal track heading inland in early winter. He followed it and found, far on the tundra, a hooded seal, dragging itself laboriously along on severely frost-bitten flippers.

This happens fairly frequently with harp seals on the Labrador coast. They are caught in a bay or fiord when fast-freezing ice closes off its entrance, and they then head inland. The people of Labrador call such unfortunate seals "crawlers," and explain their strange behavior by saying that to the short-sighted seals the dark forest and rock in the distance may appear as open water. Occasionally large numbers of harp seal may be trapped by ice in a fiord, as the white whale and narwhal are caught in *savssats* further north. In 1954 several thousand harp seals were trapped by ice near the head of Hebron Fiord, in northern Labrador, and Eskimos told me they had killed more than 2,200 of them.

Like the harp seal, the hooded seal is born under a curse: to have a superb fur, avidly desired by man. Little harp seals become the "whitecoat" of commerce, and furriers prize even more highly the "blueback" of little hooded seals. The baby hood sheds its foetal whitecoat in its mother's womb, and emerges fully moulted, with a lovely slate-blue fur on its back, a black facial mask and a creamy white underside.

Commercial exploitation of hooded seal stocks began about two centuries ago and, as with the harp seal, probably reached its peak in the nineteenth century. In 1918, 23,000 hooded seals were taken on the Front ice, but since then the kill of hoods in this area has sharply decreased. Between 1946 and 1958, an average of 5,800 hooded seals were killed on the Front.

Norwegian sealers, who take about 90 per cent of all hooded seals, killed between 1946 and 1958 an average of 75,000 of the animals per year in all hunting areas. In the "peak year," 1951, they took 145,105 hooded seals. Considering that the total population of hooded seals is estimated at between 300,000 and 500,000, this seems like a frighteningly high toll.

Since then, some measures have been taken to ease the hunting pressure on the unlucky hooded seal herds. The small number of hooded seals pupping in the Gulf of St. Lawrence is now completely protected. Since 1961 adult seals, moulting in June and July on the ice of Denmark Strait, are no longer hunted by Norweigan or Russian sealers. Only a ship from Greenland still pursues them in this region.

But on the main breeding patch, the West Ice north of Jan Mayen, no quota limits the kill and the six-week hunting season allows Russian and Norwegian sealers ample time to fill the holds of their vessels with the pelts of young and adult hoods.

The survival of large harp seal herds in the Gulf of St. Lawrence and the White Sea now seems reasonably assured, but agreements for the conservation and management of hood and harp seal herds breeding in

The beautiful pup of the hooded seal, the fur on the back a glossy blue-black, the underside silver and cream.

international waters have lagged far behind the realities of persistent "overkill." If all the concerned governments now agree, as seems likely, at least for the harp seal herds, to establish quotas based on sustained yield, then the herds should maintain their present population levels.[2] But the days when these herds numbered many millions of seals are forever gone; million-fold death stalked them too long.

Hooded seals give birth to only one pup, but this mother has adopted a second, already weaned but still hungry pup abandoned by its mother. (Facing page) The end of the pups — streaks of blood on the ice, piles of pelts and tens of thousands of little corpses. In the background a Norwegian sealing ship in the "Front" ice.

Islands of Seals

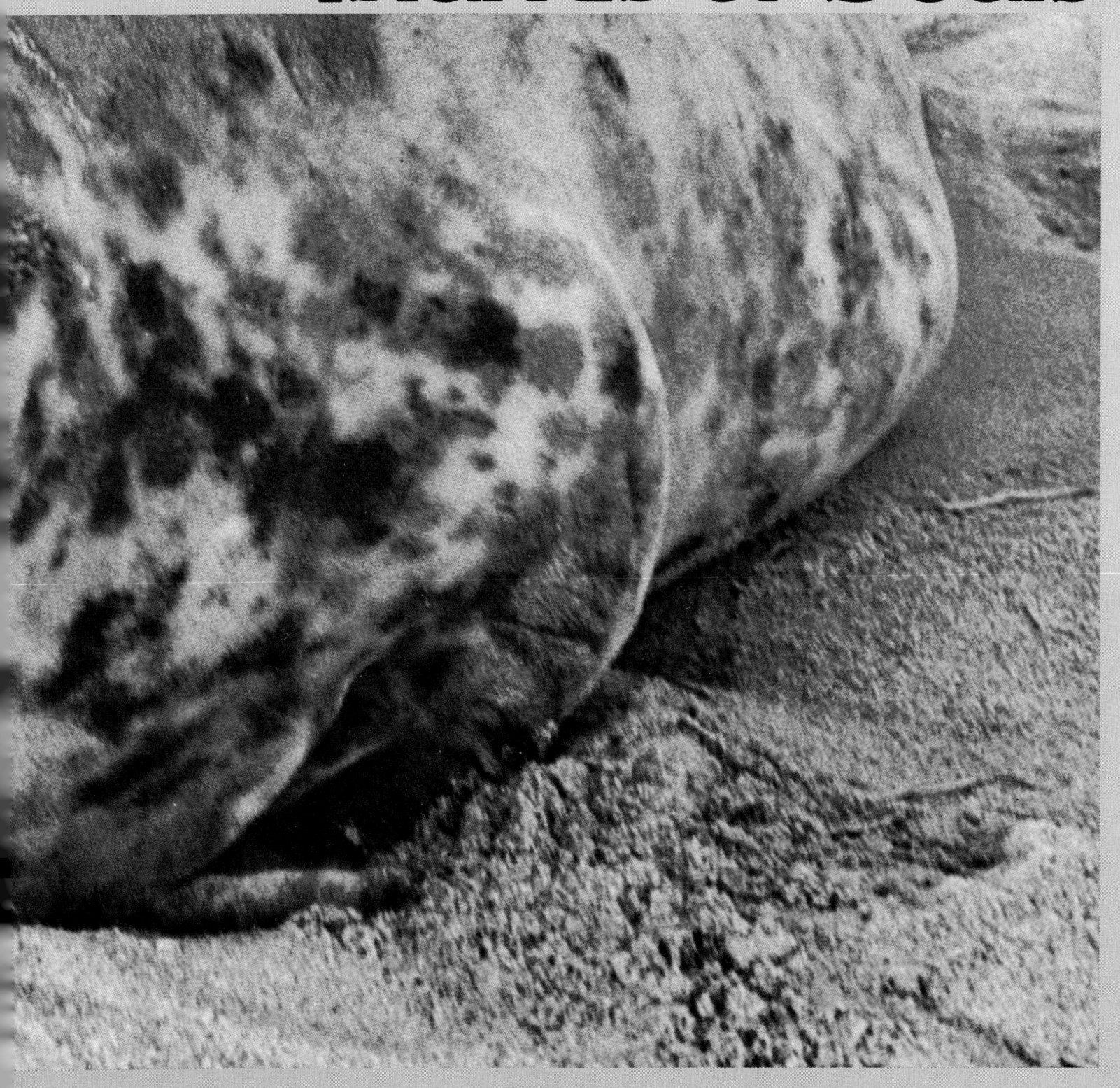

The battle of the bulls. First one lunges, then the other, as an intruder challenges the established territory of another male.

Basque Island is a tiny chip of land, two miles off Cape Breton Island on Canada's east coast. In January, when wind-blown surf pounds its rocky shores, grey seals, largest and rarest of all Canadian seals, haul out on this island to bear their young. It is a rather forbidding place for a nursery. Serrated ranks of rocky reefs surround the island on three sides. Even on calm days a heavy swell breaks violently against its shore, and frozen spray glazes much of the island in winter.

So inaccessible is Basque Island in winter, local fishermen were unaware of its seal colony until it was discovered during an aerial survey made in 1962 by Dr. Arthur Mansfield of the Fisheries Research Board of Canada.

"Be prepared to get wet," Brian Beck told me, when I asked what clothing to bring for the trip to Basque Island. Just how right he was, I found out later.

We parked our rented stationwagon near the house of a fisherman-farmer at Cape Breton Island's Michaud Point, and assembled and inflated the large, multi-chambered boat which was to take us across to Basque Island. "It's supposed to be unsinkable," said Brian.

Basque Island, two miles offshore, lay like a darkish, foam-fringed hump in the glistening sea. The outboard motor pushed our unwieldy craft across the long swells rolling in from the Atlantic, and Ingram Gidney steered it cautiously through the maze of sharp rocks near the island. The wind blew from the island and, even before we could see the seals, we were hit by the strong, fetid smell of the great bulls in rut.

Despite their impressive size (up to one thousand pounds), and their arrogant Roman noses which have earned them the local name of "horse-heads," the big bulls were rather cowardly, and humped hurriedly towards the sea as soon as we landed. Since, in their flight, they occasionally passed through the already vacated territory of rival bulls, a few of the more enterprising males attempted, *en passant,* to mate with some of the females in their path.

About four hundred grey seals inhabit the one-and-a-half-acre island. They begin to arrive early in January and most pups are born between the 15th and 25th of January. The island is ruled by the strongest grey seal bulls. Each one of these grey-black, battle-scarred veterans stakes out his territory and jealously guards it and all the females on it. There is no lack of challenges to their authority. Mateless males circle the island, casting covetous glances at the assembled females and, when opportunity seems to beckon and the harem lord sleeps, some creep ashore as stealthily as is possible for a half-ton seal. Usually they don't get very far. As soon as they trespass on occupied territory the harem master wakes up with a jolt and comes galumphing down, looping over rocks and sand like an enraged gargantuan caterpillar.

The battles are largely sham. The bulls confront each other with outstretched heads, hiss and bellow and slash at each other's necks with sharp curved canines. Usually the dominant male, the already established territorial master, wins. The intruder turns, and hitches himself *thumpety – thumpety – thumpety – thump – thump – thump* in all haste to the sea. The victor returns to his quite disinterested females, and tries to get some rest, but uneasy lies the seal with many wives. After a while, though, the defeated males seem to accept the inevitability of their single state and give up, for the present year at any rate, all attemps to conquer territory and females.

At birth the pups have skins that hang upon them like white, downy fur coats, many sizes too large. But they soon grow into their coats. Grey seal milk is the richest of any mammal's, containing 52.2 per cent fat and 11.2 per cent protein and each day the pups gain slightly more than three pounds, while their mothers, who don't leave the island during the lactation period, lose more than six pounds daily. When the pups are weaned at two or three weeks of age, their weight has increased from thirty to close to one hundred pounds.

Brian and Ingram had come to tag the pups. Had these been newly born, it might not have been too difficult, but most of them were already one or two weeks old, weighed sixty pounds or more, and objected strenuously to this operation. They snarled and bit, and scratched with flailing, sharp-clawed flippers, and the pant-legs of the seal taggers were soon in tatters. To make matters more difficult several of the females refused to leave their pups and, on the ice and compacted snow of the island, they could slither along at a surprising speed.

Considering that nearly two hundred pups were born on the small island, mortality was surprisingly low. We found only two dead pups. When we arrived at Basque Island two bald eagles had circled up majestically, white heads and tails flashing in the sun. The dead pups, and possibly scattered placentae, had presumably been their meal.

When evening came, all but forty-odd pups had been tagged. We pushed our rubber craft into the sea and returned to the mainland. The next day looked stormy. Dark clouds hung ominously over the horizon. We waited until noon, but when the weather neither improved nor deteriorated we crossed over to the island, foam-capped waves swishing against the blunt bow of our craft and dousing us with spray that froze on us into a carapace of ice.

While Brian and Ingram conscientiously tagged the last pups I photographed a fully moulted young seal which had established itself in a pool of water among the rocks near shore. It was a male, easily recognized by its dark fur, splashed with silvery spots, and its nearly solid black head and powerful muzzle. It felt quite secure in its pool and resented my presence. It lay just below the surface of the water and, every once in a while shot out with an angry growl to lunge at me, scaring me so effectively that I once slipped and nearly joined it in the pool.

Most of the other pups looked rather ragged. The long white, woolly natal fur, stained yellowish in newborn pups by amniotic fluid, was starting to moult, first near the muzzle, and then on the back, showing the short-haired dark coats of young males and the silvery grey of young females. Some pups were frantically alarmed by our presence and tried to slither away or attack. Others paid little attention, unless they were too closely approached or touched. A few, who had apparently just imbibed a massive dose of their mother's fat-rich milk, slept blissfully, short flippers pressed to bulging bellies, like bloated burghers on a beach!

The storm, which had hung in abeyance all day, decided now was a good time to blast loose. Great breakers rolled high onto Basque Island, and around the reefs the sea seethed in angry, foam-whirled turmoil. We waited until a wave rolled high onto the beach and then tried to ride out with it in our clumsy rubber craft, paddling wildly. But before we reached deep water, the next comber curled neatly over our boat and swamped it. Brian paddled, I bailed and Ingram yanked on the starting cord, while we drifted closer and closer to the reefs.

We were nearly on the reefs, when the motor caught and slowly started to push us away from this most immediate danger. Thanks to its air chambers, our boat could not sink. But since it was nearly full of water, it wasn't particularly buoyant either, and instead of riding the waves we plowed through them. Just as we passed the last reef, the propeller blade caught on a rock, the motor let out one anguished screech, then lapsed into silence, and we were drifting, slowly, out into the Atlantic.

While Ingram tried to fix the motor, Brian and I managed to bail most of the water from our wildly lurching craft. After that it rode the high waves quite nicely. Only the wave caps kept breaking over the boat,

Grey seals battle in a shallow lagoon on Sable Island. In the foreground, a female seal and pup.

covering us and everything in the boat with a layer of ice and making Ingram's efforts at reviving the recalcitrant outboard even more difficult. Finally he got it going again, but somehow most of the horsepower seemed to have been knocked out of it and it spluttered along in an off-beat, half-hearted fashion, constantly on the verge of complete collapse. Slowly, agonizingly slowly, we rode back to the mainland.

    It started to get dark, the wind increased, the waves got bigger and it looked as if we might be pushed past the last spit of land. We barely made it. Brian and I paddled flat out and, as we drifted past the point, Brian jumped into chest-deep water and hauled us to shore, just as the motor expired in a last rattling gasp. Brian produced a reasonably dry package of cigarets from a plastic bag and we sat in the snow and smoked in contemplative silence, watching the great waves thunder over the point. "I told you you'd get wet!" said Brian.

Canada's total grey seal population is estimated at about five thousand. Their breeding grounds are on islands, most of them as remote and inaccessible as Basque Island. The only one which is fairly easy to reach is Amet Island in Northumberland Strait, between Prince Edward Island and New Brunswick. Its seal population plummeted from about three hundred in 1954, to less than thirty in 1965, and local fishermen, who bear

(Left and below) Mating grey seals. The bull grabs the female by the scruff of the neck to prevent her escape. *(Right)* An angry bull rears up and roars his defiance.

(Above) A grey seal pup in its down-soft natal fur. (Right) A female with a group of pups, some partly moulted, on Basque Island.

the grey seals a grudge since they claim the seals destroy nets, were believed responsible for this drastic decline.

Other grey seal colonies are on Deadman Island (one of the Magdalen Islands) in the Gulf of St. Lawrence and on remote Sable Island in the Atlantic, 150 miles east of Halifax. Since scientists were tagging all grey seal pups at all known colonies, they were mystified by the occasional capture of untagged young animals. Grey seals are rarely hunted except along the Labrador coast and at Miquelon, a small French-owned island south of Newfoundland. There three hundred grey seals were shot in the summer of 1964. Not one bore a tag and they obvioulsy came from a hitherto undiscovered colony.

The riddle was solved in January 1965, when a helicopter pilot from a Canadian icebreaker saw a large colony of adult seals and pups on the ice of George Bay, west of Cape Breton Island. These had to be grey seals since no other seals pup at this time of the year. In 1966, Brian Beck and another technician from the Fisheries Research Board visited this colony on the ice, and found it numbered nearly a thousand grey seals.

Before the causeway linking Cape Breton Island to the rest of Nova Scotia was built, the powerful currents running through the Strait of Canso discouraged the formation of extensive ice sheets in George Bay. Since the causeway was built and the current stopped, the seals from Amet and other islands where they were hunted by fishermen had left land for the safety of the ice, just as grey seals in Europe's Gulf of Bothnia breed on the ice. The fact that young grey seals are born with a camouflage "whitecoat," making them inconspicuous on ice and snow but clearly visible on dark land, seems an indication that grey seals, in the past, normally bore their pups on ice.

In 1967 adverse winds and mild weather discouraged the formation of suitable ice fields in George Bay, and the grey seals returned again to Amet Island and a few neighboring islets in Northumberland Strait to bear their pups.

Tagging returns indicate that adult grey seals spend the whole year in the vicinity of their breeding islands, but young animals travel widely. They have been recovered from Maine, in the United States, to Cape Chidley, at the extreme north tip of Labrador. In 1962 a small grey seal colony was discovered in the United States, on Nantucket Island, off the Massachusetts coast. Archeologists have found grey seal skulls in the kitchen-middens of prehistoric Indians along the New England coast, but the colony of fifteen animals discovered on Nantucket Island is the first known in the United States in historic time.

In Europe, grey seals are more common, their total number there being estimated at about fifty thousand. They have been known, and hunted, since prehistoric times; lovely legends of singing seals were invented, and many myths grew up about this happy "people of the sea," who came

A grey seal mother snarls warningly as a human intruder comes near her pup. (Above) A kiss of recognition. A female may identify her pup by its smell.

ashore to sing and celebrate, and sometimes even married earthlings. That gentle monk of the seventh century, Cuthbert, who became the sixth bishop of Lindisfarne, loved the grey seals living on the Farne Islands off Northumberland. He would go and talk to them, and he also tamed and conversed with otters and eider ducks. Before his death he made the islands into an animal sanctuary, and so they have remained to this day. I am glad they made him a saint!

But St. Cuthbert's example was rarely followed. To the fishermen of Wales, northern Scotland and the Orkney Islands, grey seals were a source of blubber, oil, meat and pelts (the first lighthouse lamps burnt grey seal oil). By 1914 only a few hundred grey seals were left near the British Isles. Then the Grey Seals Protection Act was passed by the British government, and the seals recovered fairly rapidly to a present number of about thirty thousand. Now they are the object of a prolonged and bitter dispute between Britian's seal lovers, who want them left alone, and Britian's fishermen, who say the seals tear nets, eat salmon, and ruin the cod fishery – they would like to see the seals reduced to an absolute minimum level. While not able to accommodate such widely divergent aims, the British government has chosen a middle path: to maintain the grey seals at their present numbers by the periodic and supervised "cropping" or "culling" (two nice euphemisms for killing) of pups.

In Canada grey seals are protected during the breeding season and are rarely hunted, except at Miquelon and Labrador. Only at Miramichi Bay in New Brunswick, where they are supposed to endanger the extensive salmon fishing industry, are they actively pursued; the government pays a bounty for all grey seals taken in this area.

This is a purely local problem, however, and the number of grey seals killed is small. Much more serious is the fact that the grey seal is now known to be a vector of codworms. These parasites grow in cod (and other fish), but only reach sexual maturity when the fish is eaten by a seal. The worms' eggs are passed into the water with the seal's excrement, are eaten by fish and the chain recommences. Codworm are a bane of the fishing industry. They are not dangerous to humans, but wormy cod filets look unappetizing and are hard to sell. Many fish plants employ full-time "worm-pickers," who have the disagreeable task of removing the one- to two-inch-long worms from the fish. The loss to the fishing industry is considerable and grey seals are partly blamed for it. As a result, grey seal numbers will be reduced. In 1967 and 1968 a Nova Scotia sealing firm was allowed to "harvest" most of the pups on Basque Island, and both in 1969 and in 1970 five hundred adults were killed at various breeding places near the Canadian coast.

Canada's largest grey seal group, numbering about fifteen hundred animals, is protected by the sheer remoteness and inaccessibility of its

(Left) A moulting grey seal pup. The moult begins at the head. Then the white natal fur on the back falls out, until only a fringe remains on the sides. (Below) A peaceful group of grey seals on Basque Island — a sleepy bull, left, a female, and several pups in various stages of moulting.

breeding grounds, on fog-shrouded Sable Island far out in the Atlantic. The crescent-shaped island, roughly thirty miles long and barely two miles wide, has a grim reputation. Since its discovery in 1500 by the Portuguese more than four hundred ships have come to grief on the island's low, shifting shores, and thousands of sailors lie buried beneath its drifting sands.

The day in January when Dr. Arthur Mansfield and I approached the "graveyard of the Atlantic" in a little plane from Halifax, the feared island was a vision of loveliness. Long, bottle-green breakers, capped with crescents of gleaming foam, rolled onto its yellow beaches. Grotesquely eroded dunes wound in undulating patterns along the island's shores, and in the valleys between them, feeding on the sere and fallow grass, stood dark clusters of wild horses, their long manes and tails flowing in the wind. There was a dense group of grey seals near the island's west spit, immature animals and young bulls not yet powerful enough to claim one of the females on the breeding grounds; and a similar agglomeration at the east spit. But on the great triangle of sand, about three miles long, between the east spit and the first dunes, hundreds of grey seals lay in small groups, dark spindle-shapes on the golden sand.

After shoveling out piles of sand that had drifted in through chinks and cracks, Dr. Mansfield and I established ourselves quite comfortably in the old abandoned lighthouse keeper's home, a few miles from the

An angry female ready to charge.

island's east end. We walked each morning to the seal colony to observe the behavior of the animals from the vantage point of the last dune, or to tag the hundreds of pups.

Unlike Basque Island, where space is at a premium and territory is occupied by the most powerful bulls, who control the harems of females that live on this territory, Sable Island has plenty of space and here most of the grey seal bulls seemed to be monogamous. The seals usually lie together in family groups – a male, dark, majestic and often deeply scarred around the neck, its much smaller, sleeker female, and her pup. Most of the seals lie fairly close to the shore, but some had humped nearly a mile inland, and we even found a few family groups right on top of one of the dunes.

As on Basque Island, most of the grey seal bulls took off in a hurry the moment we approached. But some did not. They stood steadfast with their females, arched their necks high in a threat posture, growled warningly and, if this was ignored, charged with a sudden and powerful forward lunge. The aggressiveness of females seemed to vary with the age of the individual, the age of her pup and the distance to the sea. Those nearest shore nearly always abandoned their pups, undulated rapidly towards the sea and threw themselves with a last big humping motion into an oncoming wave. Young females and some with very young pups often were ready to defend their offspring valiantly, and it also seemed to me that their determination to stand and fight increased in direct ratio to their distance from the saving sea. After a while I found I could tell pretty accurately which females would attack. They seemed to adopt a special stance, head held low to the ground, eyes bulging so that some of the white showed, and emitting a persistent, fairly high-pitched gurgle. If I overstepped her limit of toleration, the female threw herself forward, mouth agape, and snarling furiously. Fortunately for us, the seals were slow on the sand, and it was easy to dodge their attacks.

One day on the beach I came across a female who ignored me completely. When I took a closer look, I realized why. She was blind. Both eyes were whitish, and only when her pup snarled at me, did she realize danger was near. She was well fed and her pup healthy and rotund. Blind grey seals in good physical condition are, in fact, no great rarity. They have been seen at many breeding colonies.

When walking out to the very end of the island, past the breeding grounds, I came frequently across young males, lying on their sides, eyes tightly closed and sound asleep. When I gave one of them a gentle nudge with my boot, he reluctantly opened his eyes, stared at me in unbelieving horror, flung himself onto his belly and thumped off in high gear, a great undulating mass of fear-propelled blubber.

At the very end of the spit lay the immature grey seals and young bulls, about 150 or 200 of them, a closely-packed, fishy-smelling coterie. As I approached them they rushed into the sea en masse, then bobbed up some twenty yards offshore, rising and falling with the waves, long lines of bullet-shaped whiskery heads, staring curiously at me with bulging eyes.

For about two weeks after the pup is born the female is not in the mood for mating, and all advances of the great bull who hovers in attendance are fiercely repelled. When his time finally comes his approach is usually direct and rather overwhelming. To prevent escape, he simply steamrollers his female, pinning her down with his thousand-pound mass and, just to be sure, with his teeth he grasps her roughly by the scruff of the neck, and clasps her firmly with his flippers.

Sable Island is famous for its fogs. Once, in recent years, the island was shrouded in thick fog for nine uninterrupted weeks. We were near the east tip of the island, late one evening, when fog banks rolled in with surprising speed. We walked back, beckoned homeward by the stabbing, milky shafts of the automatic Light.

Dark shapes of seals loomed up, vast and unreal in the swirling fog. Through the greyish pall, directionless and all-pervading, came the myriad voices of the seal colony: the hiss of an angry pup, the enraged gurgle of a female, the deep-voiced snarl of a male, the eerie forlorn wailing of a forsaken little one, all blending together with the dull roar of the breakers and the sibilant whispering of wind-driven sand into a symphony befitting this lonely island, cradle of seals and graveyard of ships.

A young grey seal bull sleeps on his side, blissfully unaware that he has unwelcome company. Awakened by a nudge, he turns over in haste and rushes off across the sand.

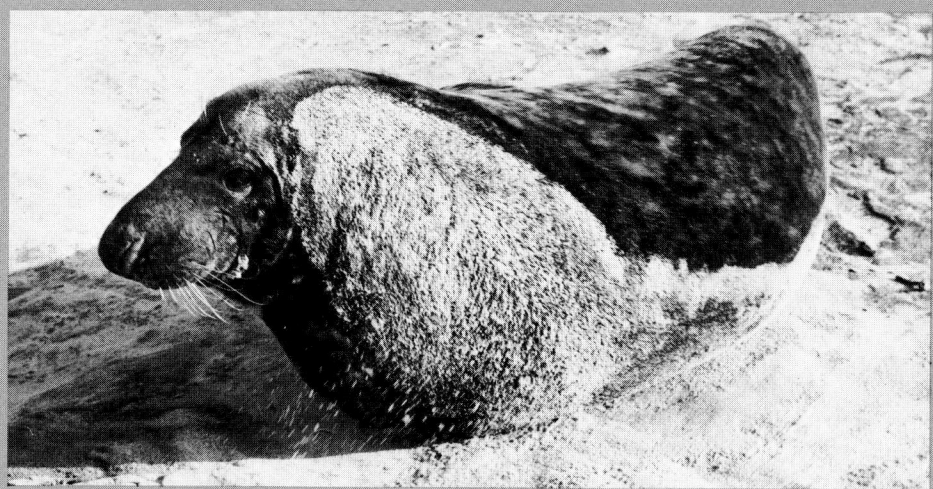

"Bestes à la grande dent"

"*Oogli tavani!*" ("the oogli is there") said Tommy Nakoolak and pointed to an islet in the distance. An *oogli* (pl. *ooglit*) is the hauling-out place of walruses. The one we were now approaching was off northeast Coats Island in northern Hudson Bay.

From a distance this nameless little island looked like the immense humped back of some giant warty whale. A closer look showed it to be covered with wall-to-wall walrus. As we circled the islet, scarcely 150 yards long by 50 yards wide, in Tommy's Peterhead boat, most of the walruses humped heavily toward the sea, oozing over the rough rock like a herd of giant tusked slugs.

Once in the water, though, there was nothing sluggish about them. They rushed around, a dense group of 150 or more walruses, dived for an instant and surfaced again with a deafening chorus of angry bellows. The females placed themselves protectively between the boat and their calves and once, when an inquisitive youngster came too close, its mother surged up, clasped it in her broad front flippers and rushed it out of harm's way.

Most of the great bulls, and some of the females without calves, remained on the island. "Can I go ashore?" I asked. Tommy looked dubious. "Maybe no good. Maybe *aivik* (walrus) attack," he said, but when his son Mark and his son-in-law Peter Alogut offered to come with me, he agreed.

We took the canoe ashore. Mark remained with it and every time a walrus came close, he slapped the side of the canoe with the palm of his hand. The sound, magnified in the water, was sufficient to scare the walrus away from our fragile craft.

The bulls refused to budge as we crept over the slimy rocks, half nauseated by the miasmic stench of the oogli, smelling like a ripe cesspool. The great walruses glared at us with bloodshot eyes, and each time we stepped close they reared up with a flash of ivory tusks. Then, as if exhausted by this sudden show of defiance, they sagged forward, heaving great wheezing sighs, blinked their eyes a few times and went to sleep.

"Stay above them," Peter warned me. Like sea lions, walruses can swing their hind flippers forward, pushing themselves up and forward at the same time with their broad, powerful flippers. They advance thus on

When the ice has melted, walrus crowd together on favorite hauling-out places, such as this islet in northern Hudson Bay. (Overleaf) A large walrus pod in the shallow water near their hauling-out place.

and Bretons were already well established on the islands, exploiting the seahorse colonies. They had even less use for English interlopers than had the walruses, and chased them off with rifle fire.

The surveyor-general Samuel Holland, who visited the island in 1765, reported that at the *échouries,* as the places were called where waluses hauled out and were killed, as many as eight hundred were slaughtered in one hunt. A herd was "cut" into groups of forty or fifty animals, and these were driven inland and butchered. At the two most important échouries more than two thousand walruses were massacred that year. Although Holland's estimate that as many as one hundred thousand walruses hauled out on Deadman Island alone is probably wildly exaggerated, some walrus pods on the Magdalen Islands are thought to have numbered eight thousand to ten thousand individuals. In another fifty years it was all over. The last walrus was reportedly killed at La Petite Echourie in 1799. Since then only the odd straggler from the north has been seen in the Gulf of St. Lawrence.

The walrus colony on Sable Island suffered the same fate. Sailors from wrecked ships killed many. The merchants of Boston sent out regular hunting expeditions. One, in 1642, returned with "... 400 pairs of sea horse teeth, which were esteemed worth £300." In addition the men brought back twelve tons of oil and many hides. When the Huguenot pastor Le Mercier of Boston tried to establish a colony of French Protestants on Sable Island in 1738, the island's walrus had already been exterminated. From time to time Sable's shifting sand uncovers a walrus skull, mute witness to their existence, long ago, on this lonely island.

Walrus products, of course, were always valuable. Hakluyt reported the skins already were prized in his time because "the leatherdressers take them to be excellent good to make light targets [shields] against the arrowes of the Savages." Also "the teeth of the sayd fishes ... are a foote & sometimes more in length: & have been sold in England to the combe & knife makers" – at a price twice as high as that fetched by elephant ivory. Finally, a good friend of Hakluyt's, Mr. Woodson "of Bristoll, an excellent Mathematician & skilful Phisition ... assured mee that he had made tryall of it [walrus tusk] ... and had found it as soveraigne against poison as any Unicornes horne."

Walruses are gregarious animals. Most of the year they spend in groups on the pack ice, floating above the clam beds on the sea floor. When they are hungry they dive, plow up clams with their tusks, and apparently use their long sensitive whiskers to sort food from the debris and to shovel it into their mouths. Since one rarely finds clam shells in their stomachs, they must somehow suck out the mollusks' most delectable parts, particularly the long siphons, and spit out the rest. Where clams are rare, they will eat whelks, crabs, sea-cucumbers or shrimps.

While some walruses stay on the oogli as the boat approaches, most abandon the island and swim around it in a noisy, surging mass.

After a good feed, the walruses return to their ice floe. They love company. So many sometimes pile onto a floe, even lying on top of one another, the ice may subside under the sheer weight of walruses.

In winter some walruses migrate to areas where currents keep the water open. In the western part of Jones Sound a large number of icebergs are usually frozen into the sea ice. Since the greater part of each berg is below water, currents tend to shift and turn them and there is a ring of upheaved broken ice around each iceberg. Here it easy for a walrus to keep a hole open, and one can see the tell-tale brown stains where they have rested near the foot of the bergs.

Other walruses settle in areas where wide leads among the ice make it easy to haul out. After feeding they lie on the ice and sleep. Occasionally they snooze too long. Shifting currents may close the lead and the unfortunate walrus finds itself locked out. Then begins a desperate search for another lead or similar area of open water. The walrus will drag its great bulk for many miles, living all the time off its accumulated fat. Eskimo hunters from Grise Fiord on Ellesmere Island told me they have followed the trail of such a wandering walrus for more than thirty miles. They have found dead walruses occasionally, and others completely exhausted and emaciated, the thick skin on flippers and underside worn to the quick.

The calves are born in spring, usually in May, and stay with their mothers for at least two years. When in danger, or tired, the calf often climbs onto its mother's neck, or is clasped by the female with her flippers and carried pressed against her chest. The female walrus is a devoted mother. She will defend her calf against all enemies and never forsake it, no matter how great the danger.

Fridtjof Nansen (who, after trying to reach the North Pole from his ship *Fram*, spent a winter on Franz Josef Land living mainly on walruses), once shot two calves in a pod. All the animals rushed into the water, except

the mothers of the killed calves. "One sniffed its litle one and nudged it, apparently unable to understand what had happened. It only saw blood streaming from its head, at which it wailed and wept like a human mother."[1] When the female pushed the calf towards the water, Nansen, afraid to lose his catch, tried to stop her. "But the mother anticipated me. She grasped the carcass of the dead cub in one of her fore-flippers and disappeared with it into the depth. The second mother repeated this manoeuvre.... I thought that the dead cubs would have to emerge to the surface but nothing was seen of them.... The walruses must have carried them along for a time." So Nansen went to another herd and shot there a calf and its mother. "It was a touching sight to see her bend over her dead young one before she was shot, and even in death she lay holding it with one foreleg."

Such devotion, thought it may move few humans, does undoubtedly have a considerable survival value for the calves, an important factor in such a slow-reproducing species. Females mate when they are five or six years old and usually have a calf only every two or three years. Apart from man, walruses seem to have few enemies. Polar bears reportedly try to kill the odd calf, but apparently rarely succeed. More dangerous may be the killer whale but, since in emergencies many adults will come to the rescue of an endangered calf or even another adult, even the fierce killer whale may think twice before joining battle with sharp-tusked, battle-proven walrus bulls.

As long as Eskimos hunted walrus from kayaks or umiaks the numbers they killed were relatively small compared to the total walrus population of former times. Such hunts were also highly dangerous. Walrus will attack a boat when pursued or pressed (although a motor boat tends to frighten them) and no skin-covered kayak or umiak was a match for their powerful tusks. Eskimos in Ivugivik told me of a man whose boat was ripped by an enraged walrus bull. Floundering helplessly in the water, the man was clasped by the walrus in its flippers and crunched against its chest with such force all his bones were broken. The Eskimos managed to retrieve the body. "It was as squashy as a sack of blubber," they said.

Even when hunting walrus from a thick-planked Peterhead, Eskimos take no chances. When I travelled with Tooma of Southampton Island through the wide leads in the late summer pack ice, he spotted a solitary walrus bull on a floe. He shot the animal, harpooned it as it surfaced and then kept firing until he was absolutely sure the walrus was dead, keeping his boat as far away from it as possible.

Once white men organized well-equipped expeditions to slaughter walruses, exploitation soon decimated the once enormous herds. White hunters were not hampered by any taboos or compunctions about killing the animals on land, and they did a most thorough job: the herds at Bear

Island, north of Norway, were wiped out; those near Spitsbergen reduced to a few dozen; and of the immense masses of walrus hauling out on the islands north of European Russia only an estimated two thousand are left. The walrus herds in the Gulf of St. Lawrence are gone and so are those on Sable Island. Once they had nearly exterminated the bowhead whale, arctic whalers turned their attention to walrus and slaughtered them by the tens of thousands. The herds of Pacific walruses fared somewhat better, but even there, in a space of twenty years, between 1860 and 1880, more than two hundred thousand were killed. Now the world population is optimistically estimated at one hundred thousand. The low estimate is that only forty-five thousand walruses are left.

In the Canadian Arctic the walrus may hold its own at the present time. Dr. Arthur Mansfield has estimated their number in the Coats Island – Southampton Island area at about three thousand, with at least another three thousand further north in Foxe Basin. Smaller concentrations of walruses are found along the coasts of Baffin Island, Devon Island and Ellesmere Island. Their total number in the Canadian Arctic may be between eight and ten thousands.

In the early 1960s, the average annual kill in the Canadian Arctic was about one thousand walruses but, as Dr. Mansfield has pointed out, to this number must be added at least another three hundred killed but lost through sinking.

Lately, though, the pressure may have eased a bit. During the summer of 1968 I visited all settlements along Ungava Bay and Hudson Strait. From here, in the past, had been launched some pretty efficient hunting expeditions to walrus areas, particularly to Akpatok Island in Ungava Bay and, further west, to Nottingham and Salisbury Islands. In the two previous years, the Eskimos hadn't bothered. They had enough seal meat to feed themselves and, since most dogs have been replaced by skidoos, they did not need walruses for dog-food.

At Coral Harbour, on Southampton Island, as more skidoos are used, the number of voracious huskies has decreased and with it the urgent need to hunt walrus as bulk dog-food. In 1960, there were about four hundred sled dogs on Southampton Island, and it was estimated the Eskimos needed 115,000 pounds of meat and 36,500 pounds of fat per year just to feed their dogs!

This positive factor may soon be offset by a new trend, which already seems responsible for the destruction and waste of large numbers of walruses in Alaskan waters: the need for ivory. Under a law passed in Canada in 1931 only Eskimos may hunt walruses, and no unworked ivory and no walrus hides may be exported from the Arctic.

In recent years the Eskimos have found that, while soapstone carvings sell well, ivory carvings sell better, and in many settlements there is a

Curious walruses tread water to stare at the photographer.

definite shortage of walrus ivory. When I visited Ivugivik, the Eskimos were busy cutting up old implements made of ivory: parts of harpoons, tips of kayak paddle blades, tool handles or the ivory necks of sealskin floats. From this ivory, aged and yellow, they carved birds and seals for "export." The temptation to go and kill walrus solely for the tusks will rise as the demand, and the prices, for ivory carvings go up.

The biggest oogli in the Canadian Arctic is probably on the east coast of Coats Island. There Dr. Mansfield counted fifteen hundred hauled-out walruses on August 8, 1961, and the next day twenty-one hundred walruses clustered chummily together on the flat sloping rocks.

I walked to the oogli on August 14. The majority of walruses had not yet arrived, or had already left, because there were only some five hundred on the promontory. I could hear them, and smell them, long before I saw them. Most lay on shelving rocks close to the sea, but some, especially big bulls, had hauled themselves quite high onto the top of a rock ridge.

Unfortunately the only side from which I could approach was with the wind, and the walruses, whose sense of smell seems to be keen, had ample warning of my presence. A few females with calves rushed to the water before I was within one hundred yards and, as I walked closer, the whole mountainous mass of walrus, after glaring worriedly at me with bulging eyes, suddenly moved, en masse. For the span of several seconds, a brownish-red avalanche of massive bodies, tusked heads held high, cascaded down the slippery slope and belly-flopped into the water, rushing over each other in their mad scramble to get away. Once they were all in the sea, panic ceased and curiosity set in. As I stayed partly hidden behind a stone, groups of twenty or thirty walruses, chiefly males, swam back and forth along the oogli, stopped suddenly and raised themselves neck-high out of the water, stared inquisitively at me, and disappeared with a splash the moment I made the tiniest movement, such as advancing the film in the camera. On land they had looked majestic but rather ugly, lumbering and ungainly, an immense mass of muscle and blubber, wrapped into a wrinkle-rippled hide.

Now in the water, looking curiously at me in the sort of bewhiskered, avuncular way of elderly gentlemen, interested but slightly disdainful, they seemed droll and charming, moving with graceful ease. Once a whole row of them, perhaps forty or more, rose in single file some forty yards from me, all staring intently. It was evening, and the last light of the sun skimmed low across the water, glinting on the long curved tusks, and I thought that of all the names given to the wondrous walrus – sea horse, sea cow, sea ox – the most appropriate is that of Champlain, who called them *bestes à la grande dent,* "beasts of the long tooth."

Nanook - The

Great White Bear

It was 6 a.m., and Brian Beck's narwhal-catching crew at Koluktoo Bay on Baffin Island — David Robb and I — sat in sullen matutinal silence, drinking coffee. Brian, more energetic, had already made breakfast and was now at the creek near our tent brushing his teeth. Suddenly we were jolted out of our glum meditations by a hoarse, inarticulate cry, and seconds later Brian burst into the tent, frothing toothpaste, and yelled: "Bears!" We dashed out just in time to see a polar bear mother and her two cubs swim out into Koluktoo Bay. Brian had been squatting near the little creek, sleepily brushing his teeth when he happened to look up and saw the bear trio, five feet away, watching him with interest.

We followed the bears by canoe. The mother swam strongly for the opposite shore of the bay, dog-paddling with her great front paws, and trailing her hind legs. The two chubby cubs, probably about eight months old, rode pickaback, their sharp-clawed paws dug into their mother's thick fur. When they slipped, the female dived and surfaced underneath them, giving the cubs a chance to get a more secure grip.

Polar bears are good swimmers, but rather slow. A man in a rowing boat can overtake them, and any motor-driven boat speedily catches up with swimming bears. In the water the mighty white bear is completely helpless. I have seen Eskimos take their boat right alongside a swimming polar bear and shoot it at point-blank range. It is not pleasant to watch, but I can understand it. The Eskimo kills in order to eat the bear, to use its pelt for clothes or, more often, to sell it. But I fail to see what "thrill" white trophy hunters get out of killing swimming bears. To shoot a bear, who can neither flee nor defend itself, at a distance of a few feet, requires no skill or courage. It only takes money.

Our female bear soon realized she could not escape us. She headed for the nearest shore, grunting nearly continuously to her cubs. On shore she was safe. Sure-footed she loped up a steep scree slope, stopping from time to time to let the cubs catch up.

The bears come from the north, travelling up Milne Inlet. Rather than make the many-mile round trip around the great Bruce Head peninsula, they had taken a short-cut across a neck of land. I have seen polar bears use similar overland routes from one sea area to another, in other parts of the Arctic. Thus, to judge by the number of tracks, they seemed thoroughly familiar with the thirty-mile valley chain making a convenient path across Devon Island, from Wellington Channel to Jones Sound. This indicates that polar bears, despite their apparently haphazard wanderings, have a good knowledge of the lay of their land.

Their realm is vast and harsh. It covers more than five million square miles of circumpolar sea and land. Polar bears have been seen hundreds of miles from land in the Arctic Basin by scientists on the United States drifting station T-3, and arctic expeditions have found their tracks (and

The beautiful massive head of a female polar bear.

those of the arctic fox) within two degrees of the North Pole. But since seals, the polar bear's main prey, are rare in the central Arctic Basin, such far-north excursions are probably rare.

Towards the south, polar bears range as far as Moosonee, at the head of James Bay, nearly at the latitude of London. Occasionally bears travel with the pack ice down the Labrador current, as passive migrants. Nearly every year one or more polar bears are seen at the north tip of Newfoundland. In 1966 one was shot at Great Brehat, six miles north of St. Anthony, and in 1967 another was killed at Ship Cove. Some years ago a polar bear was killed in the Lake St. John region of Quebec. It had presumably drifted into the Gulf of St. Lawrence with the ice and was trying to hike overland back to its northern territories.

Long ago, though, polar bears were common along the Labrador coast. The eighteenth century trader-explorer-hunter Captain George Cartwright came upon more than a dozen polar bears, scooping salmon out of a south Labrador river in the fashion of the great Alaskan brown bears. Dr. Charles Jonkel of the Canadian Wildlife Service has suggested this may have been a locally learned technique of the Labrador bears. Since this group has vanished, no further observations of polar bears catching salmon or char in this manner have been reported.

If the polar bear's range has diminished (they once regularly visited Iceland and may even have denned on its northwest coast), it may be due as much to the warming trend of the arctic climate as to overhunting. But they are not nearly as numerous now over their entire range as they were a couple of centuries ago, when arctic whalers on several occasions counted more than a hundred bears in the immediate vicinity of a whale carcass, and often shot them by the score. Between 1905 and 1909, Dundee whalers alone killed more than a thousand polar bears off the East Greenland coast.

As a result of such ruthless slaughter, the world population of polar bears is now estimated to be only between ten and twelve thousand, with about six thousand in the Canadian north. Russian estimates are even lower. Soviet scientists believe only eight to ten thousand polar bears may be left.

There are some indications that, if the polar bear's range has shrunk, that of the black bear is extending northward. When I visited Nachvak Fiord, northern Labrador, more than a hundred miles north of the treeline, in 1968, an Eskimo had just shot a young male black bear. Several black bears, including a female with cubs, were seen near Fort Chimo on Ungava Bay; and two hundred miles to the northwest, near Payne Bay, another black bear was shot in 1968, again far north of the treeline and the first the Eskimos of that region had ever seen. Still stranger is the case of the black bear killed in 1967 near Povungnituk, northeastern Hudson Bay. It was shot in midwinter (when it ought to have been hibernating) on a small island in Hudson Bay.

Polar bears are inveterate wanderers and one sometimes finds their tracks in the most unexpected places. When I crossed Spitsbergen's icecap with a Scottish expedition we were surprised to find polar bear tracks far inland at an altitude of about four thousand feet, near the base of Newtontoppen, the island's highest mountain.

Although polar bears seem to roam at random across the vastness of their arctic realm, there may be seasonal movements and possible migrations, as yet little known and understood. The Eskimos of northern Hudson Bay, for example, believe polar bears from this region travel with the pack ice in spring and summer south and east, then wander northward on foot along the west coasts of James Bay and Hudson Bay. Since such a theory is based mainly on the observation that bears appear and disappear seasonally, it is, at best, highly speculative.

It is a fact, though, that pregnant female polar bears resort to certain specific areas in early winter to dig their dens. The main known denning areas are on Franz Josef Land and Wrangel Island, north of Siberia (with about 150 dens each); Kong Karls Land, an island group east of Norwegian-owned Spitsbergen; and along the central east coast and northwest coast of Greenland. In Canada the major denning regions so far located are southern Banks Island (polar bears from this region may wander to Alaska where there are no known denning areas); eastern Southampton Island in Hudson Bay; and eastern Baffin Island. Near the now abandoned settlement of York Factory on the Manitoba coast of Hudson Bay, south of Churchill, Dr. Charles Jonkel of the Canadian Wildlife Service discovered in 1970 one of Canada's major polar bear denning areas.

There, in October or early November, each female digs a roomy lair, usually an oval chamber about ten foot long and four foot wide into a snowbank on a slope. A small vent, created by the warmth of the immured animal, admits fresh air, but temperatures in the dens may be 40 degrees higher than outside air temperatures.

The cubs are born in late November or December, usually one if it is the female's first birth, twins as a rule thereafter and, very rarely, triplets and even quadruplets. These cubs, who will grow into the largest carnivores on earth, are no bigger than a rat when they are born and weigh about $1^1/_2$ pounds. They are blind, deaf and nearly naked. They lie in their mother's warm, heavy fur and suckle her fat-rich milk. When they whimper with cold, the mother curls protectively around them to lift them even further off the icy floor of the den.

During hibernation the female is not in a torpor but merely lethargic and sleepy, dozing away the days and weeks of winter, while her cubs grow and become ever more heavily furred.

Like all bears, the polar bear can easily raise itself on its hind legs to get a better view. Balancing with outstretched front paws, it can even walk a few steps.

The Canadian scientist Richard Harington, who has made an extensive study of the polar bear's denning habits, measured the temperature of one den in Southampton Island. Then "we opened the hole wider to find out more about the occupants. A glistening black eye and twitching muzzle were instantly applied to the aperture by the mother bear. While she paced the den floor beneath us, uttering peevish grunts, we were just able to discern her two young cubs huddled against the far wall of their snow house."[1]

Females with yearling cubs and, occasionally, some male bears, may also den; but most barren females and adult males prowl the frozen vastness of the Arctic in darkness and bitter cold, searching for *agloos,* as Eskimos call the breathing holes of seals. There a polar bear lies in wait, to scoop out, with one mighty swoop of its long-clawed paw, the luckless pinniped when it comes up to breathe. But each seal has many breathing holes and a polar bear may often wait in vain. During prolonged blizzards, even males may seek shelter in temporary dens, but when the storms abate they resume their solitary search for food in the darkness of the arctic night.

Although seals are the bears' chief prey, their keen sense of smell will lead them to anything remotely edible. Two Eskimos from Ellesmere Island, with whom I made a six-week spring hunting trip, shot seven polar bears. Five had nothing in their stomachs, one stomach contained parts of a gull or fulmar and another a considerable quantity of muskox hair. Presumably the two bears had found the carcasses of these animals.

Those seven bears were quite fat, one indeed so fat it could gallop only awkwardly when pursued by the sled dogs. Yet at the same time we met a Cornwallis Island Eskimo who had shot a miserably emaciated bear. "Nanook walk much, eat little," the Eskimos said. Such bears, they explained, seem very high-legged, while a fat bear of greater weight may look chubbier and smaller. In addition to these two types of bears, Baffin Island Eskimos insist there is a "super-bear," twice as large as ordinary bears; similarly colossal bears have been reported by Siberian tribes. The sixteenth century Dutch explorer Willem Barents reported polar bears twelve and thirteen feet long from the Spitsbergen area, and a male measuring 12 foot 4 inches and weighing a reported 2,210 pounds was killed in 1960 off the Alaska coast. Normally, though, adult males weigh between six hundred and one thousand pounds, and females between five and seven hundred pounds.

In March the female digs her way out, and her cubs, now chubby and heavily furred, emerge from the murky den into the dazzling world outside. They follow their mother closely at first, but soon their natural curiosity gets the better of them and they scamper hither and yon.

Near the coast of Devon Island, I followed once the spoor of a female polar bear with two cubs. The female was in search of *nunarjaks,* the oval lairs ringed seal mothers excavate under snowdrifts as nurseries for their pups. The female's tracks showed a purposeful patrol from one frozen-in iceberg to the next, while the erratic pattern of the cubs' tracks showed they obviously had had a marvelous time, chasing a feather here and climbing an ice hummock there to slither down its side. Under a wind-fluted snowdrift in the lee of one berg, the female had scented a seal den through several feet of snow. She had scooped out great chunks of compacted snow, then reared up and, with one violent blow, broken through the icy roof of the seal's lair. Two limp, furry flippers were all that remained of the seal pup. Not much of a meal for a polar bear who can, given the chance, devour one hundred and fifty pounds of blubber at a sitting.

Female polar bears are devoted mothers and will defend their cubs with ferocious courage, but they can also be stern disciplinarians. A Grise Fiord Eskimo told me how he had watched through his telescope a female with two cubs hunt seal in spring.

This is the time seals moult and they like to lie in the spring sun, close to their breathing holes or leads, and snooze. They are fitful sleepers. Every minute or oftener a seal suddenly awakes, raises its heads and looks carefully around. If all seems safe, the seal slumps forward and sleeps again.

This polar bear parked its cubs in the shelter of a pressure ridge. Then it began the long, wary stalk towards the seal who, like all seals wishing to have a reasonable life expectancy, lay on a flat ice plain, well away from pressure ridges and similar hiding places where enemies might lurk. As soon as the seal slept, the polar bear crept ahead, pulling itself forward with its front paws and pushing with the hind paws. The instant the seal moved, the bear froze into immobility, an indistinct yellowish hump on the ice. Only its coal-black India-rubber nose stood out starkly against the surrounding whiteness. ("It is unmistakable miles away,"[2] Vilhjalmur Stefansson noted). The Eskimo said (and Peter Freuchen claims to have seen it) the polar bear hid its black nose behind a paw. As soon as the seal slept, the bear inched forward again. It was nearly within pouncing distance, when one of the cubs, no longer able to bear the suspense, came prancing out from its hiding place. The seal took one look at the running cub and dived into its hole, while the mother bear loped back and gave the offending cub a cuff that knocked it head over heels.

Spring, fortunately, is the season of plenty for polar bears when most are grimly in need of ample food. Winter is hard on the wandering males, by March the fat reserves of the lactating females are usually exhausted. Now seal pups are easy prey, and unwary seals can be caught near their breathing holes.

Regal and relaxed, a polar bear on a rock ridge on Coats Island in northern Hudson Bay.

on a lake near the Hudson Bay coast, their yellowish fur aglow in the rays of the late evening sun.

Once the ice breaks up, hunting is more difficult. Polar bears are good swimmers but seals swim much better. Then the bears travel across the drifting pack in search of seals sleeping on the floes. When they have spotted one, they noiselessly ease themselves into the water, hind legs first, and swim under water to the floe with the seal on it. At the spot closest to the seal, the bear surges out of the water and pounces on the seal, grabbing it with its claws and crunching its skull with its long canines. Unless forewarned, few of the small ringed seal can escape this lightning attack, but the large bearded seal may squirm out of the bear's horrible grip. Eskimos with whom I travelled in Frobisher Bay shot a bearded seal whose back and side bore the deep parallel lacerations of polar bear claws.

In southern Hudson Bay and James Bay, where the ice begins to break up and disappear as early as July, the bears seem to head for favorite islands or shore areas. During an aerial survey conducted by Dr. Jonkel of the Canadian Wildlife Service in the summer of 1968 more than one hundred bears were seen on the islands of James Bay. Other areas of polar bear concentrations are Cape Henrietta Maria, at the west juncture of Hudson Bay and James Bay (recently created a "polar bear park" by the Ontario government), parts of the Manitoba coast and Coats Island in Hudson Bay.

At this time the bears may switch from a wholly carnivorous to a predominantly vegetarian diet. It is rather surprising to see a polar bear, the world's largest carnivore (with the possible exception of the Alaskan brown bear) sitting peacefully on an arctic meadow munching grass and sedges. This diet may be supplemented with seaweed, sorrel and other plants, with berries in fall, and in lemming years some polar bears methodically hunt these little rodents. Sometimes they raid colonies of eider ducks or snow geese, slurping eggs and catching the odd brooding bird. In the fall of 1967, a hunter near the coast of Hudson Bay saw a polar bear stalk one of his goose decoys with patience and skill. At ten yards it pounced but when it only got a mouthful of papier mâché it flattened every decoy in sight.

When food of any kind is hard to procure, the bears seem to estivate. They dig pits into sand ridges and eskers, varying from shallow depressions to holes six foot deep, and there they doze, expending a minimum of energy. As soon as ice forms in fall, the bears probably head out to sea again to hunt seals and replenish their shrunken fat reserves before the winter comes. At this period they may congregate in considerable numbers in certain regions. From aerial surveys and ground observations, Dr. Jonkel has estimated that there were about three hundred polar bears in the Cape Churchill area of Manitoba at the beginning of November 1971.

Polar bears are closely related to the European brown bear. Both are probably decendants of a common ancestor, *Ursus etruscus* of the early Pleistocene. In captivity brown bears and polar bears mate and produce fertile offspring. As polar bears or proto-polar bears spread to the game-rich ecological niche of the Arctic, they became superbly adapted to its harsh conditions. Until man appeared, they were undisputed rulers of this vast realm. Only in the water may polar bears be in danger from killer whales or the attack by the odd irascible walrus bull.

    The Romans knew the great white bear of the Arctic and, to amuse the populace, flooded specially built arenas where polar bears were pitted against seals in aquatic battles.

    In medieval times, the odd polar bear was brought from Iceland and even Greenland to the courts of Europe. They were extremely valuable. Depending on whether its owner had mercantile or spiritual ambitions, he could trade his bear for a ship plus cargo or a bishopric.

    The depletion of polar bear stocks began in earnest with the beginning of arctic whaling and exploration. As the bowhead whale became increasingly rare, whalers augmented profits by hunting polar bears. Later, sealers did the same. Between 1893 and 1908 Norwegian sealers alone killed, on the average, 415 polar bears each year. In 1924, they killed 714.

    In Victorian times polar bear skins became a sort of status symbol, as rugs in front of the fireplace on which babies could be photographed; the demand for skins continues, nearly unabated, to the present day. Until 1970 only Eskimos and Indians in Canada were allowed to hunt polar bears; but, since prices have risen from forty dollars, and less, two decades ago to two hundred dollars, and more, now for prime skins, they have pursued the bears relentlessly, killing in recent years above six hundred each year.

    Since 1967 a quota system has limited the kill in the Northwest Territories to 383 (it was raised to 414 in 1969) but since the ordinance is difficult to enforce in some areas, there is a danger that the quota may be exceeded.

    In 1970 Canada opened its Arctic to white polar bear hunters. Under the new ordinance passed by the Northwest Territories Council each Eskimo settlement is allotted an annual polar bear quota (the total to come to 414 a year). The Eskimos can either shoot the bears themselves and sell the skins, or they can sell their rights to killing a polar bear for the hefty sum of three thousand dollars: two thousand to be paid by the white hunter to the Eskimo community which sells a bear from its quota, and one thousand to his Eskimo guide. Whether shot by whites or Eskimos, the quota for the Northwest Territories remains fixed at 414 polar bears per year.

    In Alaska sport hunting of polar bears, often by the rather unsporting method of hunting them with planes, is big business. For the "thrill" of

The polar bear dog-paddles with its front paws, letting its hind paws trail. It is a good but rather slow swimmer.

killing three hundred polar bears off the Alaska coast in 1965, well-heeled big game hunters shelled out about four hundred and fifty thousand dollars.

"Arctic safaris" leave each summer from northern Norway in small luxury yachts to shoot polar bears off the Spitsbergen coast in the water, using the "just-lean-over-the-railing-and-press-the-trigger" technique. Wintering trappers and personnel of weather stations in Spitsbergen kill more than one hundred bears each year. Henry Rudi, an old-time Spitsbergen hunter, known as *bjorne-kongen* (bear king), whom I met in Tromsö, had killed more than seven hundred polar bears!

In Greenland, too, more than one hundred bears are killed each year and, until recently, the demand by zoos for polar bears (there are about one thousand polar bears in the world's zoos) added even further pressure on the diminishing bears, since the most common method of getting cubs is to shoot their mother. Thus the total kill of polar bears still exceeds one thousand per year.

As the world's kill of polar bears increased sharply, concern for the fate of this slow-reproducing species (at best two cubs every three years) resulted in measures for its protection, management and conservation. Since 1955 polar bears have been completely protected in the Soviet Union. Only a few may be captured under licence for zoos. The zoos, too, probably feeling guilty about their contribution to the demise of polar bears in the wild, are making efforts to rear their own polar bears – a difficult project because polar bears do not, as a rule, breed readily in captivity. Yet some zoos have had success, most notably the one in Nuremberg, Germany, where 37 polar bear cubs were born between 1945 and 1960. Nineteen of them were successfully raised.

Females with cubs are now protected everywhere. Since 1967 the kill of adult bears in the Canadian Arctic has been limited by quota, and Alaska, too, has recently set a yearly quota of about three hundred and fifty bears and reduced it to three hundred polar bears in 1970. Kong Karls Land, one of the major denning areas, has been declared a polar bear preserve by the Norwegians. Closed seasons in Greenland and Alaska afford the bears some respite.

Polar bears are extremely curious and, where they are not hunted, show little fear of man. Their apparent amiability poses a special problem in the North of today. Polar bears became pets at several military installations in the North. At one station a polar bear became so tame, the men lured it regularly into their mess hall to feed and photograph it. At another station the cook fed slops to a polar bear each day. One day he slipped, fell in front of the bear – and that was the end of the cook.

Polar bears are not normally aggressive but, when they are provoked, teased, injured or scared, they may attack and, despite their placid, slow-moving appearance, their charge is lightning quick.

A polar bear relaxes on a rocky ridge. (Below) Relying on smell rather than sight, a polar bear sniffs curiously as he approaches. (Far right) Fall hunting on land is lean for polar bears; they patrol the sea coast hoping to find food.

A polar bear swimming through the glistening waters of Hudson Bay.

In the Churchill area, where bears are numerous in late fall, the great attraction for them is the garbage dump, in the vicinity of which each night up to twenty bears may be seen. Polar bear watching is a popular form of entertainment and as long as people stay in their cars it is pretty harmless. But vandals have killed some polar bears at the dump and wounded others. Somehow encounters with bears do not bring out the best in man. The bear either arouses in him the urge to kill, or he regards this powerful and potentially dangerous animal as a sort of friendly overgrown pooch.

When I was in Churchill in 1967 a polar bear had discovered a washed-up white whale carcass on the beach not far from the highway. It gorged itself on the somewhat putrid blubber and then, replete and lazy, fell asleep near the side of the road. A whole busload of school children were driven out to admire the bear. The children formed a ring around the bear, threw candy at him (which he ate) and a few stones – then one adult came so close the bear knocked a camera out of his hands. Fortunately, the bear was so sated and somnolent he couldn't be bothered to charge, and when the kids got too annoying he just ambled off.

Sometimes such meetings have a sad ending. In 1966 a polar bear attacked and mauled a Fort Churchill boy. The bear was killed and a later autopsy by Dr. Jonkel revealed someone had previously potted it with a .22. In November 1968 some Churchill boys decided to track polar bears – one came suddenly upon a bear and was killed by it.

Inevitably after such tragedies, there is a great clamor that all polar bears be killed. The establishment of oil drilling camps along the Arctic littoral and in the Hudson Bay area poses a new problem. Bears will be attracted to such camps by scattered refuse. And the men, bored in their enforced solitude, are likely to kill the bears for "sport," or will try to make pets of them, with possibly tragic results for both men and bears.

Although the layman would be well advised to give polar bears, however amiable looking, a wide berth, professional animal trainers, particularly in Russia, have succeded in making first-class circus performers out of them. At the famous Hagenbeck circus in Germany, twenty-one polar bears were once trained to pull a sled. The project was undertaken at the request of Roald Amundsen, the famous polar explorer, who hoped to use the team in the Arctic. But the bears' trainer had no yen for the polar regions, and Amundsen was a bit doubtful whether he could handle the bear team on his own. So he used huskies to reach the South Pole, and the bears performed their act at the circus.

In order to base protection and management of polar bears on sound scentific data, the Canadian Wildlife Service started in 1966 a long-range research program. Aerial censuses were taken in the Hudson Bay-James Bay region and, starting in 1967, polar bears were trapped and tagged to find out more about their movements and possible migrations.

In July Dr. Jonkel, his two assistants Ken Coldwell and Brian Knudsen, and I flew to Southampton Island in Hudson Bay. We built a great number of "cubby-sets," and baited them with kippered snacks and temptingly smelly walrus meat. A "set" consisted of two 45-gallon drums, one filled with rocks, gravel or sand, rolled together to form a V. Top and rear were barricaded with stones and driftwood, and sharpened "stepping sticks" were driven into the ground near the front of the barrels to force the bear to approach the entrance between the drums in a certain way. Between the drums, in a quadrangle outlined by more sticks, the spring-activated steel-cable snare was set, its end attached with clamps to the filled drum. The snare was covered with soft moss and grass, and the bait tossed far between the drums, at the sharp end of the V. As the bear approached, it was more or less forced to put its paw on the invitingly soft pad covering the snare. The snare flicked up, closed itself around the paw and – the thousand-pound drum served as an efficient drag.

Polar bears, for some unexplained reason, are southpaws – they use the left paw to kill seals; the twenty-one polar bears we caught always had the snare around the left foot.

Not that we were overwhelmed by polar bears at first. In nearly two months spent at Southampton Island, Coats Island and Cape Henrietta Maria we saw several bears but caught only one.

At the end of September we set up a new trap line near Cape Churchill and our luck changed immediately. During the next six weeks, we caught twenty bears.

The first were a mother and her hefty two-year-old male cub. The mother was in the trap and, like all the bears we captured, remarkably resigned to her plight. She lay on some rocks and snoozed. The cub kept her company.

As we came close, the mother growled warningly, a deep throaty growl, sometimes followed by an odd, cat-like hissing, blowing out the upper lip at the same time. Occasionally she chomped loudly, which Ken interpreted as "come closer and I'll eat you up." On the whole, though, she did not appear especially perturbed by our presence.

Dr. Jonkel prepared the drug for the immobilizing dart and shot it into the bear's neck. She snarled – surprised – but then lay down and resumed her snooze. The cub, curious about the shining object in his mother's neck plucked out the dart and played with it. The drug had been injected upon impact and the she-bear soon passed out.

This left the cub. Dr. Jonkel prepared a second dart and walked close. The cub was ready to defend his mother. He walked out in the broad-legged stance of an aggressive bear, but when the dart hit him, he let out a yowl of pained surprise and raced back to his sleeping mother, huddled close to

This beautiful female bear, nicknamed Linda, was caught in the Churchill area by a Canadian Wildlife Service team in three successive years.

her and thus fell asleep.

We measured and weighed the bears, tagged and tattooed them, pulled a tiny vestigial premolar from the older bear to determine her age, and took a small blood sample. Ken milked the she-bear. The milk tasted like cod-liver oil. By now the drug was beginning to wear off and the female lifted her head and looked at us curiously and still somewhat dazed.

While most bears seemed neither aggressive nor particularly afraid, one at least was pathetically shy. We had built a trap on the flat, mucky coastal strip near Hudson Bay. This bear, rather skinny, but still weighing more than nine hundred pounds, had pulled the rock-filled drum about a dozen yards and in his frantic struggle got into a water hole. When we arrived the white bear had turned into a muddy brown one. On seeing us, he ducked behind a small rock. It was touching to see the immense animal trying to hide its bulk behind a foot-high stone.

A few days later we caught a young female bear. Dr. Jonkel immobilized her and we were busy working on her when Ken happened to look up. Twenty feet away was another bear, watching us with concentrated interest. As Dr. Jonkel prepared a new dart, the bear ambled closer, not aggressive or angry, just intensely curious, walking in a slow, bow-legged shuffle. He was five feet away when the dart hit him in the butt. David (as we called the bear) jumped, walked a few yards away, sat down to think the situation over and peacefully passed out.

Two days later David was in one of our traps. Signs in the snow showed he had fought furiously to free himself, but when we arrived, he had apparently come to the conclusion that fighting did not help and slept calmly beside the trap. Dr. Jonkel gave him a light dose, just enough to make him drowsy. We pinned his neck down with a pole and slipped the cable off the foot. A few minutes later David, still a bit groggy, stumbled off. The next day he was in another trap. He seemed to have learned fast the futility of fighting the heavy drag. The drum had not been moved an inch. Not even the stepping sticks were upset. David had walked into the trap, eaten the bait, lain down and was now waiting for us to get him out of his predicament.

In November we shifted our trap line close to Churchill and patrolled it with a truck. Dr. Jonkel had gone to Ottawa and Ken and I were building trap sites. One day we saw a badly limping bear. We phone Dr. Jonkel. "Try and catch him and see what's wrong with his paw," he said. Hoping to play Androcles to a polar bear, we built a trap next to an old shack near the coast. The next day a bear was in it, but not the one we had meant to catch. It was Linda, a big she-bear who had made a nuisance of herself in the fall of 1966 by hanging around Churchill's rocket range, hoping for handouts. Dr. Jonkel had drugged and tagged her then, and Linda had been "deported" by helicopter. Now she was in trouble again – stuck in our trap.

"What do we do now?" we asked Dr. Jonkel. Assured the snare sat low on Linda's immense furry paw and that there was no danger of cutting off circulation, Dr. Jonkel asked us to keep the bear until he arrived with a radio collar.

Ken and I felt rather sorry for our captive bear. Linda had fought for freedom, and the snare sat tight around her leg. Now she was careful not to exceed the four-yard radius allowed by the snare cable. After a day, we drugged her and slipped the snare onto the other front paw. When Linda awoke, she was obviously puzzled, tried her left foot gingerly, then realized the snare now sat on the other paw. Not once during the next five days did she try to pull the cable. When we freed her, the noose was still quite loose around her foot.

Although bears, especially one as rotund and well-fed as Linda, can get along without food for weeks, we felt rather guilty and tried to sweeten her captivity by getting boxes of meat scraps from the Hudson's Bay Company butcher shop in Churchill. Linda accepted our offerings with obvious pleasure. After a while I got bolder and held out a long piece of fat to the bear. Linda walked slowly to the end of her tether, then took the proffered chunk carefully from my hand. Within a day, she took even small pieces from my fingertips without ever making the slightest attempt at grabbing the hand that fed her, or hooking me from underneath with her free paw.

One day Jerry Parker, a caribou expert of the Canadian Wildlife Service living in Churchill, came to watch Ken and me with our performing bear. While Jerry looked on amazed, we fed "our" bear. Linda was now used to these daily feedings. At first she had bristled and growled when we drove up in the truck. Now she stood fearless and eager, moving her sleek, sinuous neck and small elegant head from side to side, as if in anticipation of a good meal.

"Why don't you pet her," Jerry joked, since Ken and I were acting like two kids who had just received a friendly pup. It made me wonder. There was something quite irresistible about the idea of petting the great white bear. On the other hand, I knew by now that when a bear snaps he's fast about it. But Linda seemed so friendly. I fed her a few more pieces of meat then, very slowly, brought my hand down on her head. Her round little ears folded tight against her head and she watched my move with her brown eyes, growling softly. She ducked a bit when the hand touched. I gave her some more meat and petted her again. This time she hardly growled as I stroked the soft fur on her head.

A day later I was taking pictures of Linda. She seemed restless. Suddenly she rose on her hind legs towering majestically above me. She peered intently over the hut, then dropped again to all fours. I looked around to see what had aroused her curiosity. Standing on the far side of the hut was another bear, a great male.

An infuriated polar bear attacks a sled dog. Eskimos hunting polar bears use their trained dogs to hold the bear at bay.

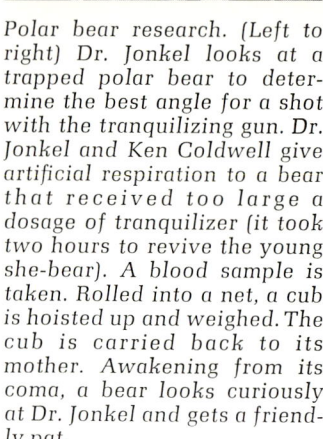

Polar bear research. (Left to right) Dr. Jonkel looks at a trapped polar bear to determine the best angle for a shot with the tranquilizing gun. Dr. Jonkel and Ken Coldwell give artificial respiration to a bear that received too large a dosage of tranquilizer (it took two hours to revive the young she-bear). A blood sample is taken. Rolled into a net, a cub is hoisted up and weighed. The cub is carried back to its mother. Awakening from its coma, a bear looks curiously at Dr. Jonkel and gets a friendly pat.

Ken and I hopped into the truck. The male, intensely curious and apparently friendly, came towards Linda. She did not like it. She growled and hissed, and when he came too close she charged. She had not forgotten her cable. She just slid to its very end, then stopped without pulling it tight. The male retreated and lay down a few yards from her.

The next day Dr. Jonkel returned. We told him about friendly Linda, but he was not amused. "Here I have been preaching to everyone to stay away from bears, and now my assistants have to go and pet one," he said bitterly. "Just how stupid can you guys get?"

When we arrived at the trap site, there were three bears: Linda in the trap and two males sleeping on a dark rock ledge some two hundred yards away. Dr. Jonkel and I walked over. The bear of the previous day ignored us. But the new bear stood up, walked slowly down the rocks and then came purposefully towards us. We backed off, Ken drove up with the truck and that scared the bear away.

We set another trap, baited it liberally, laid a trail of meat scraps leading to it and drove a few hundred feet away. The ill-humored bear picked up the meat, marched into the trap, and the snare flicked up. For a moment he was so preoccupied with food, he did not realize what had happened. When he wanted to walk away, the snare tightened and then he fought in berserk fury. He tore planks off the hut and chewed them to splinters, threw himself backward with such force he made double somersaults, ripped up shrubs and roared and raged.

Linda, only a dozen feet away, watched his mad performance with smug curiosity, completely unmoved by all the row. She did not even get up. The other male, too, had ambled over and watched the tremendous struggle for freedom with mild curiosity. He was more interested in the bits of meat still lying around.

Dr. Jonkel quickly drugged the bear and later we caught his chum; but despite this nasty experience both insisted on keeping us company while we worked with Linda, becoming so importunate in their curiosity, we had to shoo them off with the truck.

A few of the bears we caught at Cape Churchill were somewhat emaciated, and had extremely long fur on the undersides of their paws, indications that they had moved little and possibly slept through part of the food-poor summer. Some of the bears caught near Churchill were well fed but grimy, and it seemed these were "garbage bears," who paid regular visits to the dump.

Once, near Churchill, we caught a big male, a huge powerfully built bear, not friendly and complacent like Linda, but rather shy and cowed. He chomped warningly when we approached and snarled when the dart hit him. Instead of sticking, the dart bounced out and lay next to bear's head. We thought the drug had probably been injected, or at least part of it,

but the bear did not pass out. He just seemed sleepy. Rather than try a second dart, Dr. Jonkel decided to give the bear a "booster shot" with a hand syringe. "You keep him busy at the front," he suggested. I offered the bear a long bone, hanging onto it with all my strength. He took it willingly, and while we wrestled for its possession, Dr. Jonkel tiptoed up from the rear, jabbed in the needle and pressed the plunger. The bear blinked in surprise, but he was more interested in the bone, and pulled it out of my hand with a final firm tug. After a while he passed out. When we retrieved the dart, we saw it had not fired. All the drug was in it. The bear with whom I had played tug-of-war, and to whom Dr. Jonkel had just given a hand injection, had not been drugged at all. Just amiable.

But there is no relying on this apparent amiability; and in any case we were dealing with animals whose radius of movement was severely circumscribed by the length of the steel cable. To approach a free bear with any trust in its seeming good nature would be asking for severe trouble.

In a place like Churchill, where the garbage dump is a great attraction for polar bears who naturally congregate in this area in late fall, bear trouble can crop up at any time. One evening I had supper at Jerry Parker's house. Suddenly the phone rang. It was his neighbor. "I just thought I'd let you know, a polar bear is raiding your garbage cans," he said. We turned on the porch lights. Outside stood a bear, fishing tidbits out of an overturned garbage can. Had one of us unknowingly walked out, he might have bumped into the bear. Not far away, two Indians were severely mauled by a bear one winter evening when they came upon him by chance.

In 1967 and 1968 Dr. Jonkel caught a total of sixty bears in the James Bay-Hudson Bay region. Over twenty of these bears were subsequently recaptured by Dr. Jonkel in the same area. Until early 1969 not one of the marked bears had been killed by a native hunter. By the spring of 1970, Dr. Jonkel had caught a total of 161 bears, 121 of them captured for the first time. Only one of the marked bears had been killed by an Eskimo hunter, 280 miles north of where it had been tagged. Most of the recaptured bears had not moved far from the area in which they had been previously tagged.

Among the bears recaptured in the late fall of 1968 in the Churchill area was our old friend Linda. Bigger now but not much wiser than when first marked in 1966, or caught in 1967, she had again stumbled into a trap. She was still amiable and friendly, but this time no one stroked her head. It was better so. Polar bears should not be petted. They should be admired, preferably from a distance; as monarchs of the Arctic, roaming its infinite vastness of rock and snow and ice, superbly adapted to the rigors of their realm.

(Above) Dr. Jonkel attaches a tag to a bear's ear. (Below) A shy polar bear hides behind his paw. (Facing page) An amiable trapped bear accepts a piece of meat from Ken Coldwell.

Footprints in the snow — the spoor of a polar bear.

The Little White Fox

The stillness of night lay on the land. Snow covered the immense rolling tundra, and rock ridges and eskers broke through its white mantle like dark, gaunt ribs. Blue ice, candled into myriad silvery rods, still locked the tundra lakes, but they were rimmed already with hems of sable water. It was the quiet of winter's last days; the hush of a white, frozen world before the spectacular entry of spring – and color and life.

Suddenly this vast and awesome silence was broken by a high-pitched, querulous yap. Three dark dots seemed to glide across the surface of the snow – and stopped; the little fox barked again, then trotted away noiselessly on small furry feet, like a white wraith in the luminous opal light of the arctic night.

We saw the foxes quite often from then on. They had their lair amidst a mass of rocks, and in June we could hear the yipping and deep-down growling as the little foxes fought over food. The parents had long since moulted their long, down-soft, pure-white winter fur, for grey-brown, yellow-brindled summer pelts. The kits were hungry and the parents busy procuring food. Occasionally they raided our larder of caribou meat, and it was funny to see a little seven-pound fox grimly determined to make off with a twenty-pound haunch. At other times they stalked lemmings or voles in the marshy valley near our camp. The fox walked along stealthily, stopped and tensed as it heard the chittering of a rodent, bounced forward in an arched bound and snapped it up.

One day Charles Dauphiné, Jerry Parker and I hid near the lair, hoping to get pictures of the arctic foxes. We waited for an hour, then Jerry and I got fed up and walked to a nearby lake to look for an old-squaw nest. We had hardly gone two hundred yards, when one fox slipped out of the den, ran to within a couple of feet of Charles who had remained behind. It stopped in front of him and looked at him, uttering plaintive mewing, gull-like sounds, then ran back to its lair, curled up on a rock above it, and went to sleep, but slid back into its hole when Jerry and I returned.

Most dens are dug into the sides of sandy eskers, into mounds of sandy clay or sometimes into the sides of river banks. They can be quite complex subterranean excavations. Russian scientists have found dens with more than forty exits, although only a few of these were kept in good condition, while the others were overgrown.

There was a great fox den near one of our polar bear traps at Cape Churchill. It occupied nearly an entire hillock, overgrown with bushes, and was well-provided with a network of exits and entrances. From them the foxes ventured forth to pinch the bait from our bear trap. They were quite blatant about it. One day I had just replaced the bait and as I turned to go back to camp, a fox was already dragging it out. Our bait at this trap site was strictly for the foxes and we never did catch a bear there.

Near the den entrances, foxes liberally strew scats and food remnants. These decompose and provide nutrients for a luxuriant vegetation. This makes fox dens fairly easy to spot, even from a plane: lush little oases of bright green in a generally more somber landscape.

Newly born arctic foxes are blind, and hardly bigger than mice. There are usually six to eight in a litter, but numbers can vary from four or fewer to twenty and more, depending upon the availability of food. This rise and fall in fox numbers, most scientists agree, is due to the cyclic occurrence of their main prey, the lemming.

A lemming looks like a tail-less, chunky mouse. Actually it has a tail, but so tiny it is hardly visible. The ears are small and rounded, and generally the lemming is a prime example of what is called Allen's rule. Since heat loss is proportionate to the extent of body surface, arctic animals preserve body heat by a reduced surface size. They tend to be chubby and compact, and the size of ears, tails, snouts and legs is diminished to a minimum.

There are several species of lemming. Most of them wear the same colored coat at all seasons, but the varying lemming of the northernmost regions changes to white in winter. They are spunky, irascible, active little creatures. I've kept them often, but never long.

For one thing, in a camp the only containers available for a captive lemming are usually of metal, and in these they will make a racket that is quite remarkable for an animal so small. And they are so non-stop active that, for fear of heart attacks, I always release them quite soon. I admire their courage. A cornered lemming sits back on its haunches, gnashes its yellowish teeth, twitters in nearly frantic fury and bites most efficiently the hand that tries to grab it.

In summer they tunnel through the tundra tussocks or dig burrows into the ground. Their homes must be quite tidy, since they use special outdoor toilets where they deposit their droppings in neat little piles. When two males meet, they start a wild wrestling match, squeaking shrilly and biting each other in the cheeks. In winter lemmings build globular nests of dry grass in the snow and line them with feathers, or muskox wool when it is

A blue-phase arctic fox in the Thule district of Greenland. The blue phase is predominant where the foxes live primarily on seabirds. Where the foxes live mainly on lemmings, the white phase is predominant.

available. From these nests they drive networks of long tunnels underneath the snow, chomping off all the grass in their path. Elton, in his marvellous book *Voles, Mice and Lemmings*,[1] calls them "fat busy, agile mowing-machines."

Lemmings are most famous for their population cycles. Females in captivity have produced as many as sixteen litters in a year. Their daughters become nubile a mere thirty days after birth and start procreating. Within another sixty days or so the granddaughters are in the reproduction game, and soon numbers become astronomical. In the wild, females rarely have more than five or six litters in a good year, and only the offspring from the first litter may breed that same year. But, given hundreds of millions of lemming mothers, even that situation can produce a fantastic population explosion.

One year there are hardly any lemmings. The next year there are a few more. The third year they are plentiful and the fourth year they can be so numerous that, in the words of one observer, "it is difficult to walk without stepping on lemmings." Then they may start those famous mass migrations

The alert arctic fox must be a resourceful scavenger, especially when lemmings and ptarmigan are scarce.

that have startled men for centuries. Pontoppidan, the eighteenth century bishop of Bergen in Norway wrote: "At these times they gather in great flocks together . . . like the host of God, to execute His will, i.e. to punish the neighbouring inhabitants by destroying the seed, corn and grass; for where this flock advances, they make a visible pathway on the earth or ground, cutting off all that is green; and this they have power and strength to do till they reach their appointed bound, which is the sea, in which they swim a little about, and then sink and drown."

Such years are feast years for foxes, and for all those other animals of the north that prey heavily on lemmings. Foxes have larger litters and raise them without trouble. Jaegers lay some extra eggs, while in lemming-less years they may not lay any eggs at all. The great arctic owl has plenty to feed its downy young and all will mature, while in years of lemming dearth only the strongest owlet of the whole clutch may survive while its siblings starve.

When lemmings are numerous, they probably form the bulk of the arctic fox's diet. When lemming are few the four-year fox cycle, too, goes on the wane. Foxes may kill their young for whom they have not enough food, and the adults roam endlessly in search of things edible. In 1922, a year without lemmings, hordes of hungry foxes descended the Labrador coast. One was shot on Cape Breton Island, Nova Scotia, more than five hundred miles south of its arctic home range.

Summer, though, is not too bad, even when lemmings are scarce. There are nests to plunder – foxes love eggs – and birds to catch, although the ptarmigan, another important prey, is also a cyclic animal; if both lemmings and ptarmigan fail, the little arctic fox falls on hard times.

When food resources permit, the arctic fox tries to be provident and lays up caches of killed animals and of eggs to help him eke out the grim winter months ahead. The Danish scientist Alwin Pedersen once found such a store in Greenland, consisting of thirty-six little auks, two young murres and four snow buntings, as well as a large number of auk eggs. The eggs weer heaped into a little pile and the birds had been laid out in a neat row. Pedersen estimated the cache contained enough food to last a fox at least a month.

Despite such foresight, winter may be grim for the little foxes. Those on the tundra trot noiselessly across the snow, listening for the betraying twitter and squeak of lemmings scurrying along in their tunnels below. When a fox has located a lemming, he bounces high into the air and comes down stiff-legged, tail held high, to break through the snow's crust and pin down the rotund rodent. The foxes also try to dig out lemmings, scraping away with such fervor the snow flies as from a miniature snowblower but, since lemmings have such an extensive network of tunnels, they often escape.

Hard pressed by hunger inland, many foxes move towards the frozen sea. Some follow the polar bear, as the jackal follows the lion, hoping for leftovers. When a bear is hungry and has caught a seal he will devour it – blubber, guts and meat – and leave little for his hungry retinue. But when the hunt is good polar bears are fussy feeders. They strip a carcass of blubber, their favorite food, and the foxes have a feast.

Whenever in the Arctic one fiinds the great plantigrade spoor of a polar bear one is nearly sure to find near it the dainty little tracks of arctic foxes. Bears wander endlessly, and the foxes follow, to the very Pole if their shaggy provider is bound that way. Peary on his polar trek saw bear tracks at 86 degrees north, and at 87 degrees north the fresh tracks of an arctic fox. More recently, members of the 1968 Plaisted polar expedition noticed fox tracks almost daily all the way to the North Pole, but they never saw the foxes. Strangely enough, they did not see any polar bears or their tracks during the entire trip from the north tip of Ellesmere Island to the Pole.

Along the shores of the arctic seas a keen-scented fox may always hope to find something edible. A dead bird may lie in the snow, crustaceans or dead fish may have been pushed up by scouring ice blocks along the coast. And occasionally there is a blubbery bonanza, the carcass of a dead whale or walrus. This they may have to share with the Arctic's master scavenger, the polar bear; but there is plenty for all and hundreds of foxes may congregate to partake of the feast. Foxes have been seen to eat tunnels into the corpse of a sixty-foot Greenland whale, until the colassal carcass resembled a Swiss cheese.

Often, its keen sense of smell will lead the fox to another, deadly, food, the bait of a trap. As the little fox moves forward to eat the alluring bait, it steps on the release, the steel jaws snap shut and hold its crushed foot in a vise-like grip. Occasionally a fox, caught low on the foot will wrench itself free, leaving perhaps some toes in the trap. A few will tear off their foot. But most are inexorably caught. They pull and twist, and bite the icy steel until their teeth are broken to splinters, and slowly they weaken and die. Most trap lines are long. They cannot be checked very often – maybe once every two weeks, maybe only once a month – and then the little corpses, frozen in their agony of death, are removed.

If the fox is alive, it stares at its captor with glowing amber eyes, half fearful and half resigned, and the trapper steps on it, crushing its chest, so as not to damage the valuable fur. In Canada, between 1950 and 1960, nearly 400,000 arctic foxes were trapped and traded, with a total value of $5,065,937. The top year was 1954/1955, when 81,783 arctic foxes were taken. During these same ten years the price for arctic fox pelts fluctuated from a low of $8 to more than $24 per pelt. Since fox numbers wane and wax, as a rule, in four-year cycles, this periodicity is reflected in the fur return graphs, which resemble the fever curve of a malaria patient.

Before the white man came the Eskimos had little use for fox fur. They made trim out of it, children's clothes, ornamental parkas or underwear. The skins tear easily, and for day-to-day hard wear caribou and seal clothing were preferred. To catch foxes the Eskimo built ingenious stone traps, remarkably similar in type over vast regions of the Arctic. One was the *udlisau,* a six- or seven-foot-tall, beehive-shaped stone structure, with a two-foot opening at the top. Into this opening a flat stone was inserted so it could swivel. Sometimes a piece of polished ice was used instead. Just beyond the stone or ice sheet a chunk of meat was suspended as bait. When a fox climbed the udlisau and stepped on the swivel-stone to reach the bait, it tipped forward and dumped him into the trap. Or he stepped on the polished ice and slid in.

Its thick white fur keeps the arctic fox warm in the coldest weather. It also means death to many foxes, since man covets their lovely pelts.

Even more common was the *pudlati*, a small rectangular box made of heavy stones. I have found them from Ellesmere Island in the high Arctic to the Belcher Islands in Hudson Bay. Above the entrance hovered a flat stone, fitted into grooves on either side. From it a thong led to a chunk of bait in the rear of the trap. When a fox crawled in and pulled at the bait, the entrance rock came crashing down, like the portcullis of a medieval castle, and the fox was imprisoned.

The arctic fox is a dimorphic animal, that is, it occurs in two color phases – white and blue. Actually the white fox is only white in winter (the summer coat, as noted, being a modest grey-brown, yellow-brindled). The blue fox is a sort of smoky blue in winter, and a somewhat darker blue-grey in summer. The blue phase predominates in some coastal areas, particularly in regions where foxes live near bird cliffs. Thus on the island of Jan Mayen, east of Greenland, more than 90 per cent of the foxes are of the blue phase, and on the islands in the Bering Sea blue foxes also predominate, while in the Canadian Arctic they constitute less than 1 per cent of the total arctic fox population. Since the coastal blue foxes are less subject to the feast-famine cycle of lemming availability than the white foxes, their populations do not seem to have such strong fluctuations. The color phases are genetically determined. Two white foxes may have blue cubs, or two blue foxes white cubs, and if a white and blue fox mate, their litter may consist of cubs all white, or all blue, or of both colors.

Recently, a cousin and competitor of the little arctic fox has invaded its realm. The much bigger red fox is on the march to the north. A study made by Dr. A. H. Macpherson shows the red fox reached southern Baffin Island in 1918. In 1943 it was caught in central Baffin Island, by 1948 it had reached Arctic Bay in the northwest of the island; and in 1962, Akpaleeapik, a Grise Fiord Eskimo with whom I made extensive trips, captured a red fox, an old male, on Ellesmere, Canada's northernmost island. Despite persecution and competition, the little arctic fox perseveres. There are years when it seems rare and many dens remain empty. Then its cycle swings upward, like that of the lemming on which it preys, and suddenly foxes are plentiful again, inquisitive, mischievous, and trusting where they are not too severely pursued.

When we lived in a hut at Cape Churchill while catching and tagging polar bears, it did not take the foxes of the area long to divine in us a potential source of food. They rarely came in daytime. But in the evening the commotion started. They yapped and yipped, and barked and mewed, trying to drag off massive chunks of the evil-smelling walrus meat we used as bear bait, or clattered through our can-filled garbage pit. When we walked out with a flashlight, they stopped and stared, greenish eyes glowing in the dark, then scurried away. If we turned off the torch and sat quietly on the porch, they came back, and scampered all about us, like grey furry goblins in the light of the moon.

"Ill shapen beast"

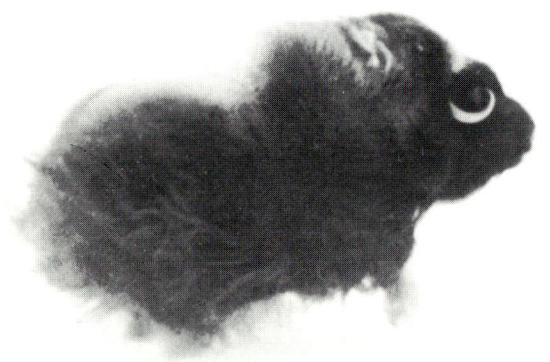

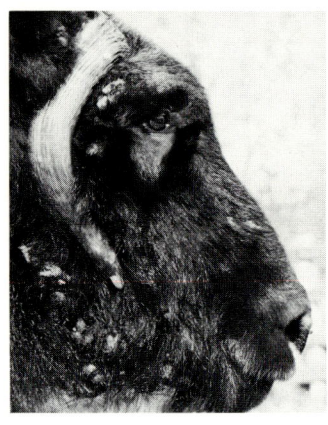My first meeting with the mighty muskox was most inauspicious. I was on the island of Spitsbergen, north of Norway. In 1929 seventeen young muskoxen, captured in Greenland, were released on Spitsbergen. They multiplied nicely until warring and food-short parties of Allies and Germans shot most of them in World War II. Since then, the survivors had increased again to over a hundred. In Longyearbyen, Spitsbergen's main settlement, I asked a man where to find muskoxen. "There's a herd somewhere in Adventdalen," he said vaguely. "Do they attack?" I asked. "The bull may charge," the man told me. "But don't worry. He's only bluffing. He stops a few feet from you." Later I found out that my kind counsellor had never been near a muskox.

Adventdalen is a big valley and I wandered about for a week looking for the herd. I explored neighboring Sassendalen, but the muskoxen weren't there either. I saw foxes and ptarmigan and ducks and geese, and several times the small Spitsbergen caribou, carrying an immense spread of antlers, as odd on this little animal as the great antlers of a wapiti bull would look on a domestic calf. And then one day as I sat on a boulder smoking and enjoying the beauty of the vast valley I suddenly saw the muskoxen, a peaceful group of some fifteen animals on a little triangular plateau whose sides fell away abruptly in steep gravel slopes to ravines below.

I climbed one of these slopes. As I appeared on the plateau the muskoxen stood up, snorted and bunched into their "hedgehog" formation, a tight semicircle of adults shielding the calves and immature animals, their heads with great curved horns facing the enemy. At the center, and a bit in front of the herd, stood the lead bull, a majestic animal, its dark guard hairs hanging nearly to the ground, its mane grizzled and the tips of the mighty up-curved horns glowing yellowish in the sun. I took some pictures, moved closer, and took some more. At fifty feet, the bull got annoyed.

He pawed the ground and rubbed his head against his foreleg. "He's only bluffing," I reminded myself. I took some more pictures and advanced. The bull snorted loudly, making hooking motions with his head. It looked unpleasantly like the warm-up for a goring session. At thirty feet the bull charged. With eight hundred pounds of enraged and sharp-horned muskox thundering at me from a very short distance I did not wait to see whether, by any chance, he might be kidding. I wheeled, raced across the plateau, jumped over the edge and went helter-skelter down the slope. When I regained my feet the muskox bull was glowering down at me, standing in superb silhouette at the edge of the plateau.

The next time I met muskoxen, they were standing peacefully in a pasture in Vermont eating apples from the hand of their keeper, Professor John J. Teal, junior, of the University of Alaska, who has succeeded in turning the muskox into a domestic animal. After studying the muskox in

Massive and archaic, a survivor from the Pleistocene, the muskox in ages past roamed the tundra of Europe, Asia and North America in company with the mammoth. This one is a bull. The cow (facing page) lacks the heavy boss across the brow.

the wild, Teal captured, with Canadian government permission, seven calves in 1954 and 1955 in the Thelon River region and brought them to his six-hundred-acre farm in the mountains of Vermont.

The calves took easily to captivity. They romped round their spacious fenced-in pasture and soon became completely tame, regarding Teal as a sort of foster-mother. How much they viewed him as one of their own, became apparent when they tried to include him in their defensive circle each time a dog (whom they instinctively feared) came into their enclosure. Their favorite game was "King of the Castle." One would get up on a little knoll and the others vied with each other to butt the "King" off his perch.

One of these calves grew into a majestic 1,400-pound bull, the biggest muskox on record. To perfect his butting technique, he had the habit of knocking down the wooden fence posts. Teal replaced them with bigger ones, but that did not faze the bull. He backed up a hundred feet or so, lowered his head, charged and Pow! there was another post in splinters. Finally Teal got fed up and built a big concrete pillar. The bull charged,

there was a resounding bang as his boss hit concrete and, stunned, he sagged back on his haunches. After a minute he got up and walked away in a daze. From then on he left all posts alone.

In the wild, muskox calves are born in spring and stay with their mothers during the summer and all the next winter, so that the cows have at best only one calf every second year. In captivity Teal weaned the calves before the rutting season in fall, with the result that his domestic muskox cows bore a calf every year.

When I visited him in Vermont his ten-year study program had come to an end, and most of his muskoxen had been given to zoos. Only two cows and a hefty two-year-old bull remained in the great enclosure. Teal had to leave. "Take all the pictures you want," he said in parting. "The two cows are gentle. The bull may charge you. Don't worry, though. He's only bluffing. He'll stop a couple of feet from you."

That, somehow, sounded ominously familiar. The cows were indeed most friendly – too friendly, really. Each time I focussed, they ambled up to see whether they couldn't coax another apple out of me. The bull ignored me for a while then, suddenly, charged, coming at me full tilt. This time there was no ravine to jump into and the fence was pretty far, so I decided to stand my ground. Three feet in front of me the bull put on the brakes, slid to a stiff-legged stop, and then came nuzzling up to mooch some apples.

Teal subsequently captured twenty-three muskox calves on Nunivak Island, off the Alaska coast. On this island thirty-three Greenland muskoxen were released in the 1930s and thirty years later they had multiplied to more than seven hundred. The captured calves were brought to a large paddock near Fairbanks, to form the breeding stock for Alaska's future herds of domestic muskoxen. By the summer of 1968 they had increased to forty-six; within a couple of decades their descendants may number in the thousands.

In 1967 Teal caught fourteen muskox calves near Eureka on Ellesmere Island to establish a second breeding herd, this one at Old Fort Chimo, near where the Koksoak River flows into Ungava Bay. *Umingmaqautik* – "the place of the *umingmak*" (the bearded ones) – the Eskimos call the muskox ranch, administered by the *Direction Générale du Nouveau Québec*. When I visited the ranch the calves, now one-and-a-half years old, were grazing placidly on a large fenced-in pasture.

Some already weighed more than three hundred pounds, and in another year they would be ready to mate, at least a year younger than muskoxen in the wild. They were tame and curious. Often when I walked into the paddock to take pictures of one muskox another came up to watch or to give me a friendly little nudge. The animals had been dehorned so that, grown up, they would not injure each other or their keepers. This made them look rather ovine, like long-haired ewes in dark coats and white

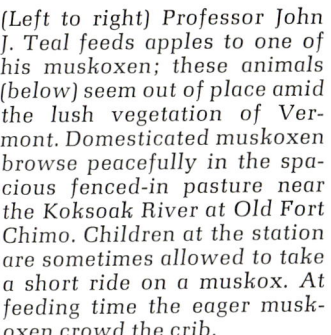

(Left to right) Professor John J. Teal feeds apples to one of his muskoxen; these animals (below) seem out of place amid the lush vegetation of Vermont. Domesticated muskoxen browse peacefully in the spacious fenced-in pasture near the Koksoak River at Old Fort Chimo. Children at the station are sometimes allowed to take a short ride on a muskox. At feeding time the eager muskoxen crowd the crib.

stockings. At night they were rounded up and driven into a high-fenced enclosure to be safe from marauding wolves and sled-dogs. Like children at bed-time the muskoxen suddenly remembered they were hungry and nibbled grass with pretended urgency; one balky little bull had to be pushed every step of the way. In May 1971 seven calves were born at the Quebec Muskox ranch.

The muskox's chief attraction for its domesticators is its wool. Underneath that immense shaggy coat of coarse guard hair the animals carry a thick, soft, silky layer of underwool, called *qiviut* by the Eskimos. It is probably the best wool produced by any animal, and one pound of qiviut can be spun into a forty-strand thread twenty-five miles long. Vilhjalmur Stefansson noted that muskox wool does not shrink when it is boiled, and it easily takes any dye. Qiviut is as light as it is soft and a sweater made from it will keep its wearer warm in sub-zero weather.

There is one hitch, though. Muskoxen cannot be shorn like sheep. Instead the wool is lifted off. Muskoxen begin to shed their underwool in May and June. It comes off in sheets and works its way through the stiff outer guard hairs. At this time muskoxen tend to look rather motheaten and ragged, trailing long streamers of qiviut. They rub against boulders and cliffs to rid themselves of the brownish wool, and great tufts of it are tumbled across the tundra by the wind.

With the domesticated muskox the wool is simply lifted off the animal as it comes loose, and pulled through the guard hairs. The muskoxen, whose skin is itchy at this time, love to be scratched and petted, and willingly part with the wool. And while a cashmere goat produces only six ounces of pashm a year, about six pounds of qiviut can be lifted off an adult muskox each summer. The calves at Old Fort Chimo shed an average of slightly more than three pounds of wool each, when they were only a year old, and raw qiviut sells for thirty-five to fifty dollars per pound, while a sweater of qiviut may cost two hundred dollars or more.

As the number of muskoxen increases in the main herds, branch herds will be established in Alaska and Quebec. Eventually herds may be raised in any other suitable part of the Arctic, from Alaska through Canada, Greenland, Iceland, arctic Norway and Finland to northern European Russia and Siberia – a land area large enough to provide forage for hundreds of thousands, perhaps millions, of muskoxen, man's newest domestic animal and potentially the world's champion wool producer.

If this comes true it couldn't have happened to a more unlikely animal. Muskoxen probably evolved in northern Asia and came to North America via the land bridge that once connected the two continents before they were separated by Bering Strait. During the Pleistocene muskoxen roamed much of Europe. Their fossils have been found in England and Germany, and in France as far south as the Pyrenees. In North America, they browsed in the

(Left) When muskoxen feel threatened, they form the "hedgehog" formation, standing shoulder to shoulder, sharp-horned heads turned towards the enemy. *(Right)* Two solitary muskoxen on a mountainside.

Snorting, pawing and rubbing the head against a foreleg are signs of excitation in a provoked and angry muskox bull. With his four-inch-thick boss and sharp up-curved horns, a bull can put a polar bear to rout.

New York region and west into Iowa. It was the age of giants. Mighty mastodons stomped across steppe and taiga; saber-toothed tigers with six-inch fangs stalked giant beavers; the megatherium, a witless sloth heavier than an elephant, stood two storeys high when it reached up to munch leaves from trees; and mammoths roamed the tundra in such numbers that half the world's ivory comes from their giant fossil tusks.

They all vanished, only the muskox persevered. As the glaciers retreated northward, so did the muskox. In Europe he became extinct; possibly early man helped him over the edge. In Siberia muskoxen may have survived until two thousand years ago and in northern Alaska they only disappeared in the last century.

The first white man to describe muskoxen was Henry Kelsey of the Hudson's Bay Company. While exploring the region north of the Churchill River and west of Hudson Bay he spotted on June 17, 1689, "Two buffilo ... ill shapen beast Their Body being bigger than an ox ... their Horns not growing like other Beast but Joyn together upon their forehead and so come down ye side of their head and turn up till ye tips be even wth ye Buts. Their Hair is near a foot long."

Muskoxen were numerous then. Nearly a century later Samuel Hearne, on his trek from Churchill to the Arctic Ocean, frequently saw "many herds of them in the course of a day's walk, and some of those herds did not contain less than eighty or an hundred head.... They delight in the most stony and mountainous part of the barren ground ... [and] they climb the rocks with great ease and agility."[1] The meat from one muskox equalled that of three caribou, the Indians informed Hearne.

The Eskimos have always hunted the muskox. The meat has been compared to beef, although Hearne said, "The flesh of bulls both smells and taste so strong of musk, as to render it very disagreeable." From the animal's long guard hairs the Eskimos made mosquito nets. The large, long-haired pelts were used as covers on the sleeping platform, and out of the keratinous sheath of the great horns the Eskimos carved ladles, flexible prongs for their fishing leisters, and toggles for dog harnesses.

Apart from man, the muskox's principal enemy is the wolf. When threatened by wolves the animals of a herd crowd together into the hedgehog formation, described earlier. Calves and young animals are kept in the center of this circle or semicircle, while all adult animals turn their sharp-horned heads towards the attacker. Should an incautious wolf approach too close, a muskox will rush out, try to gore and trample it and then return to its place in the defensive circle. This stratagem works well against wolves, but in encounters with men and rifles it is suicidal.

The Eskimos with their primitive weapons killed relatively few muskoxen. But when white bowhead whalers invaded the Arctic, they shot thousands of muskoxen to provision their ships, or equipped Eskimos with

guns to hunt muskoxen for them. Once the bison had been nearly exterminated, muskox robes, much in demand as marvellously warm sled blankets, became a valuable trading item. They were worth up to fifty dollars each at the beginning of this century, and between 1862 and 1916 the Hudson's Bay Company alone bought more than fifteen thousand of these pelts.

Arctic expeditions often depended largely on muskoxen for survival and success. Between 1880 and 1917 American and Norwegian explorers killed more than one thousand muskoxen on Ellsemere Island. They gunned down whole herds, and although many expedition members were enthusiastic hunters, they found such butchery revolting. The Norwegian Otto Sverdrup, who between 1898 and 1902 explored large portions of the Queen Elizabeth Islands, noted in his diary, after shooting a group of twenty muskoxen, that he felt like a criminal to kill "a herd of defenceless animals which had set themselves up as targets."[2]

In this century muskox calves were in great demand by zoos and also for a number of attempts, some successful and some not, to transplant muskoxen to other areas of the north. The method of capturing calves was simple and brutal. All adults of a herd were shot to get the calves, and it has been estimated that more than two thousand muskoxen were slaughtered in this manner. Somewhat belatedly, zoo directors realized they were contributing to the destruction of an already threatened animal and decided to buy no more muskox calves, a resolve made easier by the fact that under zoo conditions muskoxen were notoriously short-lived.

By 1917 muskoxen had become so rare the Canadian government ordered their complete protection. Of the Barren Ground muskoxen (*Ovibos moschatus moschatus*) on the North American mainland only an estimated five hundred survived. The northern subspecies of the arctic islands and Greenland (*Ovibos moschatus wardi*) had also been severely reduced in numbers.

Since then the animal has made a gradual comeback. The Barren Ground muskox has tripled its numbers in fifty years to an estimated fifteen hundred, and in the Thelon Game Sanctuary of the central Barrens they may be again as numerous as in 1900, when the explorer J. W. Tyrrell saw in this area "numerous bands of muskoxen feeding on the luxuriant grass or sleeping on the river bank."[3]

Ever since its discovery scientists have been puzzled by the muskox and not quite sure how to fit this ungainly ungulate into the scheme of living things. They finally compromised by giving it the rather equivocal name *Ovibos moschatus* – "the musky sheep-ox" – a bit misleading since it is neither ovine nor bovine nor, according to most observers who have come

within smelling distance, musky, except for bulls in rut. Its closest living relative is the takin, which inhabits mountainous regions in Tibet, China and Burma.

Most muskox calves are born in April or May. Temperatures at this time can be 20 and even 30 degrees below zero, but such a frosty welcome seems to have little effect upon the calves. They wear a thick, curly, dark-brown coat of wool and when afraid or chilly may stand underneath their mothers, whose long hair then hides them completely, hanging all around them like a thick curtain. They thrive on their mother's milk, rich in fat, protein and lactose, and within a week or two they begin to nibble plants.

With their immense double-layered coat muskoxen look massive and their movements are in keeping with this appearance, slow and deliberate. But when charging or in flight they can move with amazing speed and agility. I went once to Cape Sparbo on Devon Island with Paulassee, an Eskimo from Grise Fiord. We discovered a small herd of muskoxen on a grassy slope. The moment they saw us they bunched as if to face us, then moved apart in apparent indecision and began to shuffle and wheel as muskoxen often do before they flee. We made up their minds, by turning loose a dog. The muskoxen instantly closed ranks and, as the dog approached, the lead bull charged, head held low, with such speed and adroitness, the dog barely managed to jump aside and come racing back to us, while the bull returned to his herd in stately, measured steps, his great hair skirt swaying in rhythm.

We tied the dog up out of sight and sat down some fifty feet from the herd. Gradually the muskoxen quieted down. I approached slowly. Each time the bull showed annoyance, rubbing his head against a rock and snorting, I stopped for a while. Some of the cows lay down and placidly chewed their cud. In an hour I had approached to within fifteen feet. As long as I made no abrupt motions, the muskoxen did not seem particularly perturbed. From time to time the bull gave a perfunctory snort but he seemed bored and slightly blasé. This peaceful tableau changed abruptly when the dog got loose. The muskoxen rushed together, the bull made angry, abrupt hooking motions and charged.

A few miles farther we found another, larger herd. It faced us, became frightened when the dog approached from the side, wheeled in panic, fled and regrouped a few hundred yards farther away. In their sudden flight they left a little calf behind. It was an adorable animal, a miniature edition of its elders, lacking only their long overcoat of guard hairs. These only grow as a calf approaches its second winter. I walked to the calf, and it came at me on wobbly legs, valiantly trying to butt. I picked it up and carried it to the herd.

It fought at first and bleated shrilly like a lost lamb, but then it relaxed and lay quietly in my arms. The herd had calmed down somewhat, and even

(Left) The muskox calf and the herd from which it became separated in a sudden flight. Covered in a thick, curly, dark-brown coat with creamy white stockings, the little calf (right) stands unsteadily on wobbly legs — he tried to butt when approached too closely.

the lead bull (a majestic animal with nearly black fur and a magnificent sweep of horns), who had been quite bellicose at first, now seemed more pacific. Still, I didn't feel happy. In that group was a bereft muskox mother and I couldn't guess what she might do. The Eskimos had told me some gruesome tales of sled dogs who had got too close to muskoxen and were gored, eviscerated and trampled, and these thoughts didn't cheer me up. I walked a few feet, stopped and waited, and walked on again. At forty feet I put the calf down, pointed it towards the herd and gave it a little encouraging push, but it just stood there on unsteady legs, looking pathetically forlorn. So I picked it up again, while it bleated plaintively, and walked on, sweating profusely. At twenty feet I put the calf down again and respectfully backed away from the herd. We left, and watched the herd from a distance. After a while it moved forward, still in a tight group, encircled the calf and slowly walked on in stately, ponderous procession.

During summer, the muskoxen replenish their fat reserves, depleted by the long, hard winter. They browse methodically on grasses, sedges and other plants, and are particularly fond of the small arctic willow.

Fall is the rutting season, and the battle of the bulls for possession of the herds is primitive and impressive, rather like the jousting of medieval knights. When a bull approaches a herd, its lead bull will march out to meet him. At twenty or thirty paces from each other they halt. Suddenly, as upon a signal, the eight-hundred-pound bulls charge, smashing their heavily bossed heads together with a crack that can be heard a mile away. Then they return to their prior places, may nibble some grass in feigned indifference, and suddenly they charge again. They keep this up until one contestant, dizzy and weakened, flees.

Bulls not powerful enough to win a herd, or older bulls defeated and replaced by stronger rivals, become solitary wanderers; although Eskimos say in times of danger they will join, and help to defend, the nearest herd.

At Cape Sparbo I nearly bumped into a single bull sleeping behind a boulder. He took off, and another bull, who must have been nearby, joined him. Both headed straight for a herd a couple of miles away and, when we approached, formed part of its defensive circle. Later, though, they left the herd and climbed a steep scree slope. I thought surely they must slip and fall as loose rocks came clattering down. But the two bulls kept on, slow and sure-footed, until they reached the sheer cliff. There they took up their position on a small platform, backs against the rock wall, facing downward. I climbed after them, and they did not budge. Theirs was an impregnable position and they seemed to know it. They did not even go through the movements of anger, threat and excitation. They just let me approach (to within a fairly respectful distance on that treacherous slope!) and stood stolid, staring down at me in a calm and superior manner.

No arctic animal has a coat as thick as the muskox. It seems to make him impervious to cold. Muskoxen range to the very north tip of Ellesmere Island where, for nearly five months, from October to March, the sun does not rise. The mean January temperature is 30 degrees F. below zero and February isn't much warmer. Temperatures can drop to nearly 70 degrees F. below zero and storms last for days. Then even the male polar bear digs a temporary den, the arctic fox crawls into a cleft and the ptarmigan burrows deep into the sheltering snow. But the muskoxen can't hide. On the contrary, rather than seek the shelter of valleys, muskoxen prefer the exposed, wind-lashed slopes where shallow snow cover makes it easier to reach the scant vegetation. Here, too, sudden fall frost may quick-freeze the grass, preserving its nutrients, while vegetation in the valleys may have had time to wilt and wither.

In a storm muskoxen crowd together, hindquarters against the wind, sheltering the calves within this furry rampart. In the worst storms the bulls face the wind, and shield the rest of the herd – drawn up in triangular formation of which they form the apex – with their bodies. They will stand like this as long as the gale lasts, sometimes for days, never lying down, like a mighty furry bulwark.

The strong and the young survive winter, lean yet alive, but hardship, cold and lack of food take their toll of older, weaker animals. Travelling with Eskimos on Viks Fiord, northwestern Devon Island, we found a dead,

emaciated muskox cow, frozen hard as rock, though she couldn't have been dead long since no foxes, bears or wolves had yet discovered the carcass. That same spring Grise Fiord Eskimos found two dead muskoxen on Ellesmere Island, both old animals, and I found another one on Cape Sparbo, already half eaten by wolves; but Paulassee, who examined the animal carefully, thought it had not been killed by wolves but had simply died of fatigue and old age. Once new grasses sprout in May or early June, the worst is over, the calves are born and the cycle of life begins anew.

Muskoven have been completely protected in Canada since 1917, and in the three areas to which they were successfully transplanted – Dovre Valley in Norway, Spitsbergen and Nunivak Island – since their introduction; and in Greenland they have been partially protected since 1951. In the spring of 1967, however, Canada's Northwest Territories Council decided to permit the shooting of thirty-two muskoxen annually by sportsmen, despite the objections of competent biologists. It was also pointed out that muskoxen are only now beginning to spread westwards to some of the areas where they had become extinct, such as Victoria Island. But to the law-makers it seemed a temptingly lucrative proposition and the price tag, indeed, was high: four thousand dollars per muskox. Applications came from as far away as Europe. Public opposition and some scathing press comments scotched the plan, especially since even hunters agree that to march up to a stolid muskox and gun it down could hardly be called sporting. As Vilhjalmur Stefansson put it so neatly: "the word 'sport' has a curious meaning when applied to killing muskox.... I would say that equally good sport could be secured with far less trouble and expense by paying some farmer for the permission of going into his pasture and killing his cows."[4]

Thus the muskox, shaggy, craggy and archaic, is still safe in its frigid realm, slowly increasing its numbers, braving the worst weather the Arctic can offer, while its domesticated cousins, ever more numerous, furnish man with qiviut, the finest wool in the world.

To me the most memorable picture of muskoxen is one I never took. Paulassee and I had wandered all over Cape Sparbo, searching for the cave-like "house" in which the explorer Dr. Frederick A. Cook spent a winter (living, incidentally, mainly on muskox). We never found it, but we scared a herd of muskoxen and they moved away from us towards the mountains. We thought we might stop them and ran along in a depression, parallel to the herd and shielded from it.

But the herd was well spooked and kept going, first in a group then, upon an old raised beach, single file. Long fur flowing in the wind, the massive horned heads lowered, they walked along with slow stately steps, massive silhouettes against a brooding wintry sky, like a mighty procession from ages long past.

The Living Barrens

A famous thirty-volume encyclopedia gives the Barren Grounds just six lines, five to tell where they are, the sixth to say: "It consists largely of swamps, lakes and bare rock." This is a rather cavalier way of describing part of Canada nearly twice the size of France. In any case, Barren Grounds, or Barren Lands, is a brazen misnomer, for whatever else they may be, barren the Barrens are not. They are teeming with life.

Our camp was near one of the myriad nameless lakes that dot and streak the tundra, in the rugged, rocky region of the caribou calving grounds. It was early June. Snow still covered much of the land, but on warm days melt water murmured down to rivers and lakes in a shimmering lacework of brooks and rills.

The ptarmigan were busy courting, the males still in immaculate winter white, their females already in brownish-grey summer camouflage. If one is edible and defenceless, as is a ptarmigan or arctic hare, it is quite advantageous to be white in winter. Experiments carried out in Poland showed the proverbially sharp-eyed lynx could spot a dark hare on snow at three hundred yards, yet did not see a white hare on snow until it was within twenty-five yards. Charles Dauphiné, a Canadian Wildlife Service biologist at our camp, one day flushed a covey of ptarmigan. At that instant a peregrine falcon flew high overhead. The ptarmigan dived into the nearest snow bank, and became instantly invisible.

Now the cock was strutting around his female, taking small mincing steps with his elegantly feathered feet, the serrated wattles above his eyes glowing in brilliant vermilion. He fanned out his tail, and tripped along with drooping wings. The hen paid scant attention. She seemed far more interested in snipping off methodically the little green shoots poking up below the mantle of last year's withered vegetation.

Later, she had her nest not farther than 100 yards from our camp. The male sat on a nearby rock, still brilliantly white, singing his grating love song, as melodious as the creak of a rusty gate hinge. We looked for the nest among the dwarf willow bushes, but we never found it. It was too well hidden.

Nature, it seems, has assigned the cock ptarmigan a sacrificial role. While the female in drab summer dress sits quietly on the nest, the cock, white and extremely conspicuous, perches on a nearby rock, an inviting target for any foraging falcon or fox. Studies have shown predators at this time take primarily male ptarmigan. Thus die the proud little cocks, while the hens raise undisturbed the next generation of ptarmigan.

Our ptarmigan cock was more fortunate. He spent June uneaten and then he, too, changed into protective summer plumage. In between, however, he had a nasty encounter with an arctic groundsquirrel. It waddled into the ptarmigan's territory one day, a short-legged, overstuffed furry sausage, probably a male in search of a new home since its female

The siksik, or arctic groundsquirrel, eats voraciously during the summer months, doubling its weight before the long hibernation.

was expecting young and had expelled him from their cozy nest deep in a sandy esker. The ptarmigan cock was furious. He flew at the groundsquirrel and knocked it head over heels with his short powerful wings. The *siksik,* as Eskimos call the groundsquirrel, screamed in outraged anger but it wasn't halfway through its tirade when another buffet sent it spinning and it galloped off, with the ptarmigan, half flying and half hopping, in hot pursuit.

The groundsquirrel is one of the few arctic mammals to hibernate. In September it builds a warm nest at the bottom of its burrow, lugs in a good supply of food, sometimes as much as four pounds, and munches away until heavy frost sets in in October. Then the siksik curls up and goes to sleep. Its body temperature drops from 98 degrees close to the freezing point, the heart contracts slowly every two or three minutes, and while it normally breathes sixty times a minute, it now breathes only once or twice each minute. The siksik, so active in summer, now keeps the body motor barely idling, just short of stalling, using up a minimum of energy during its torpor-like eight-months sleep.

Siksiks are irascible, curious little fellows with vituperative, strident voices that can be heard a mile away. When they are excited the tiny tail flicks up and down, and each time we came past their burrows we were loudly cussed. Their den may have a couple of dozen entrances. Sometimes I set up the camera on the tripod and focussed on the hole into which a siksik had just disappeared. The next instant he would pop up in a neighboring entrance, make a few nasty remarks and zip down again. I'd shift and refocus, and there he was two yards to the left, with his cheeky chatter, sitting in front of his den and fixing me with dark-shiny button eyes.

One day I met a very fat male siksik. He dashed off in a porcine gallop, bobbing his bulbous little buttocks, and shot underneath a big flat stone. "Aha," I thought, "now I've got you!" I spent a good hour barricading with rocks every possible exit but one, while the siksik below commented on my labors with a steady stream of raucous invective. I set up the camera and focussed it on the only hole left open, sure now to get some pictures. For a while the siksik kept up his muffled maledictions; then it became very quiet. I waited and waited, and suddenly I saw the siksik. He had somehow squeezed out, and was now behind me busily nibbling at my camera bag!

Shaking off after so many months the icy shackles of winter, the tundra suddenly burst into a profusion of life. As the snow melted, the land looked somber at first in its mantle of last year's vegetation, all russet, brown and fallow. Then a green sheen spread over the "arctic prairie," as the tender shoots of grasses and sedges pushed up. Whole acres were covered with the tiny bells of arctic white heather. On the slopes the white and gold mountain avens burst into bloom, and in marshy hollows glowed the purple flowers of Lapland rhododendron.

(Above) A young siksik emerges cautiously from its burrow. (Left) An angry siksik complains in a voice that can be heard a mile away.

Elegant horned larks spiralled high into the clear arctic air, then glided earthward on stiff wings, trilling their song, as delicate and delightful as the distant tinkle of sleigh-bells. Little Lapland longspurs seemed to have a nest in every tenth tundra tussock. The males, with black heads and chestnut-colored patches on the nape, arched into the sky and warbled their gay little song as they gently floated towards the ground on set wings.

Snow buntings flitted over the rocks like feathered flakes, searching for clefts and hollows in which to build their nests. They arrive early in the north. On a bleak, blustery, snow-shrouded day in May on Devon Island I was once thrilled to hear, in the wintry gloom, the gay, triumphant spring song of a little bunting. They nest as far north as there is land and some even venture beyond it, out above the forbidding mass of shifting ice in the Central Polar Basin. There a snow bunting visited the ice-gripped *Fram,* Fridtjof Nansen's ship, at nearly 85 degrees north on May 22, 1895. "It fluttered around the ship, twittering, for some time, and then flew off towards the north."[1]

The Barren Ground caribou, for which we were waiting, spend the winter in Canada's northern forest belt, wandering about in small groups, scraping away the deep snow with broad, sharp-edged hoofs to get at the lichen on the forest floor below. To this habit they owe their name, "caribou" being derived from a Micmac Indian word meaning "the shoveler."

In April a great restlessness overcomes the animals. Small groups join and begin to move northward, picking up other herds along the way. The trickle of animals becomes a stream, and the stream a torrent, an immense host, all possessed by the same powerful urge to move north, ever north, over lakes and rivers and rugged rock ridges, travelling ever onward at about thirty miles each day, the cows large with calf far out in the van. *La foule,* "the horde," or "host," early French-Canadian voyageurs called the vast mass of migrating caribou, heading north to the Barrens in herds that once numbered hundreds of thousands of animals.

When J. W. Tyrrell of the Canadian Geological Survey made, in 1893, "an exploratory survey through the great mysterious region of terra incognita commonly known as the Barren Lands . . . of almost this entire territory less was known than of the remotest districts of 'Darkest Africa'."[2] Tyrrell saw the great caribou herds in the area of the upper Dubawnt River: "The valleys and hillsides for miles appeared to be moving masses of reindeer [caribou]. To estimate their number would be impossible. They could only be reckoned in acres or square miles." When he walked into the path of the herd, he caused "little more alarm than one would by walking through a herd of cattle in a field." The vast throng parted and flowed past him, as a river streams past a boulder.

A siksik holds a bit of root in its agile forepaws.

More than a century earlier, Samuel Hearne, travelling overland from Churchill on Hudson Bay to the Arctic Ocean, was awed by the number of caribou killed by the Indians. "The great destruction which is made of the deer [caribou] in those parts [the Barrens] at this season of the year [in fall, when furs are best for clothing], is almost incredible; and as they are never known to have more than one young one at a time, it is wonderful they do not become scarce; but so far is this from being the case, that the oldest Northern Indian in all their tribe will affirm that the deer are as plentiful now as they ever have been."[3]

The writer Ernest Thompson Seton, who visited the "Arctic Prairies" in 1907, thought the caribou "may number over thirty million, and may be double that... there is no reason to fear in any degree a repetition of the Buffalo slaughter that disgraced the plains of the United States."[4]

Unfortunately, Seton, awed by this land of "unlimited space with unlimited wild herds," seriously overestimated the number of caribou, and underestimated the white man's destructive influence even in these remote regions. Basing their opinion on the reports of early travellers and the food productivity of the caribou range, scientists now think there may have been about three million caribou a century ago. If their numbers increased further, the range could no longer support them. Animals weakened and died, or were killed by wolves, until the balance between caribou and forage was re-established.

To the Indians of the taiga – the northern forest belt – and to the Eskimos of the Barrens, the caribou was life. Its meat and fat sustained them; its hide covered kayaks and tents; spears were tipped with caribou antler; and warm clothing was sewn with thread of caribou sinew from the pelts of animals killed in the fall. They herded them into pounds; they speared them in the water at ancient crossing places; they erected alignments of *inukshuks* (stone cairns built to resemble men in silhouette) to guide the herds towards defiles where they might be killed; and they constructed circular stone ambushes which one still finds in the path of millenary migration routes. Each family needed about one hundred and fifty caribou a year, for itself and its dogs, but so vast were the herds and so scant and scattered the human population that the caribou they killed were but a drop of that living and life-giving flood.

When the white man and his firearms invaded the north the ancient balance between the caribou and its predators, whether wolf or primitive hunter, was upset. Wintering whalers killed tens of thousands of caribou, or equipped Eskimos with guns and ammunition to slaughter the animals for them. The ships that spent the winter at Herschel Island, west of the Mackenzie River delta, required as much as three hundred thousand pounds of caribou meat a year, and the whalers took only saddles and haunches, leaving the rest for the Eskimos and their dogs.

In late June or early July the ptarmigan cock (top left) begins to moult into summer camouflage. The female ptarmigan (top right and facing page) is conspicuous when perched on a rock but merges with her surroundings when pressed against the ground. (Below) A lively snow bunting male with an insect he has caught.

With the advent of traders the economy of Indians and Eskimos changed. Now they needed fox pelts to pay for the white man's goods. In early times each family had only three or four dogs. To patrol long trap lines they now required ten or more. Sometimes caribou were gunned down and left as fox bait.

Perhaps as devastating as this wholesale slaughter were the forest fires, often started by careless prospectors, that turned thousands of square miles of caribou range into black sterile desert. Lichen, the caribou's main winter food, grow slowly. After a fire it may take twenty-five years before they carpet the forest floor again in silvery-green, and more than a century if the humus has been destroyed. Caribou do not like to cross an extensive burn, and whole herds may be kept from good browsing regions beyond.

In the 1930s some biologists warned that the caribou were on the decrease. Few believed them, and the slaughter went on unimpeded. Between 1920 and 1950 from one hundred and fifty to two hundred thousand caribou were killed each year. Then in the late 1940s and early

(Above) A female snow bunting. (Below) A ptarmigan cock, still in winter plumage, brilliantly conspicuous against the background of dark tundra.

1950s disaster struck. The caribou, harassed and decimated, forsook their ancient migration routes. The great tide of animals that gave life to man on the Barrens ceased, and the Eskimos waited in vain. Many starved. The survivors fled to Baker Lake, the only settlement in the Barrens, or to the coastal villages. Today the vast Barrens is a land without people.

In the years between 1948 and 1950 the Canadian Wildlife Service carried out the first large-scale caribou census. The result was frightening. Of the estimated three million caribou a century ago, 670,000 were left. Even more ominous was the re-count of 1955. Now only 278,000 caribou remained, and by 1960 it was estimated that the herds had dwindled to 200,000 or perhaps even less.

This, it is to be hoped, was the year of the caribou's nadir. Since then they are on the increase again and may now number three hundred thousand or more. Several factors favored this increase. Eskimos and Indians, now resettled in villages, were less dependent on caribou meat than in former days, and human kill has, in fact, been sharply reduced. (Only Indians and Eskimos may hunt caribou in Canada, but there is considerable pressure to allow sport hunting for others.) Hunts in recent years have often been organized on a communal basis, with every effort made to shoot only bulls. And the native's voracious huskies have been largely replaced by motor toboggans.

Equally important, climatic conditions have favored the caribou in recent years. Most calves (as much as 80 per cent of a herd) are born in mid-June within four or five days. They are hardy creatures, but a heavy blizzard or, worse, rain followed by severe frost during these critical days can kill most of them. In good years calves constitute 25 per cent of the herds. In bad years they may amount to less than 5 per cent. And recent years have been good.

While the main herds spread out over the tundra, the pregnant cows seek out the high, remote and rough terrain of the calving grounds; the first arrived near our camp early in June after a march of some eight hundred miles. At first they still clustered in groups of five or ten but, as the time approached for each cow to bear her calf, they scattered singly into the rocky country a few miles from our camp, and we walked for days on end trying to find and tag the calves.

Dr. Andrew Macpherson actually saw a calf born. The cow was lying down. At the last moment she stood up, expelled the calf with one violent contraction, and it landed with a plunk on the rocky ground! Despite this rude arrival into a hard and chilly world the calf was up on wobbly legs in a short time and began to nurse.

The calves were lovely: the coat reddish-brown, the muzzle black, the underside a creamy white. Some were lighter in color, a soft fawn, and once we saw a nearly sand-yellow calf. If we surprised them shortly after birth,

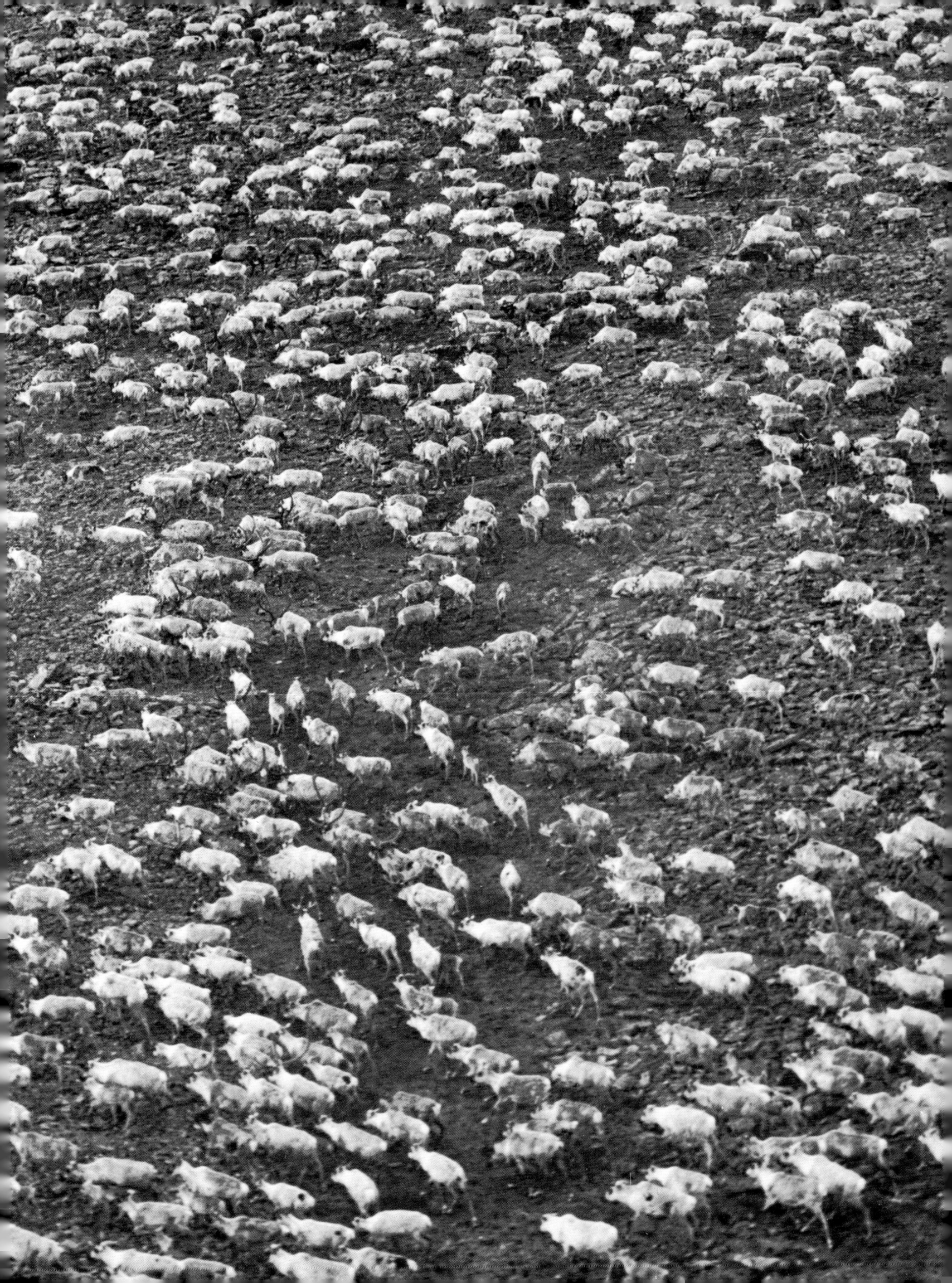

(Preceding pages) Caribou on the move. (This page and facing) Caribou does and their fawns leave the calving grounds to rejoin the main herds.

the doe fled while the calf pressed itself to the ground, not even peeping at us from its large, lustrous eyes, shaded demurely by long curved lashes. Standing, the calves looked awkward and gangly, but two or three days after birth they could run faster than we. It was humiliating. We'd spot a cow and calf, sneak up like Indians on the warpath and pounce. The doe would pirouette in a sudden, frightened jump, and gallop off, hoofs flailing, all bunched up, and then break into a magnificent long-strided trot, circling around us and the calf. The calf would take off on thin, spindly legs, running sure-footed over rock and grass while we wheezed behind, dripping sweat. If the calf was a day or two old, we might catch it. It would run and run, and suddenly lie down and could then be approached and tagged; usually it remained in its place after we left until its mother returned. But calves older than two or three days outdistanced us with apparent ease.

After a week or two the dispersed cows and their new-born calves began to leave the high grounds to rejoin the main herds. Charles Dauphiné and I spotted the beginning of this trek and walked across the crunchy candled ice of a lake to intercept the caribou at a narrows. We lay behind a boulder and watched them: a long procession of cows, their long winter fur bleached to grey-white, and ragged, walking along in a steady long-strided gait while the calves pranced around, ran away a bit and dashed back to their mothers.

When caribou walk or run, their foot joints make a loud and very distinctive clicking noise. They also cough and grunt a great deal, and a large herd on the move is extremely noisy. A biologist, who found himself in the path of fifty thousand caribou, wrote: "The clacking of their hoofs, the constant blatting of the fawns, the grunting of the females, the constant coughing and wheezing all made a roar that was deafening.'[5]

After the calving season, the great migratory urge seems to have spent itself and the caribou's movements during the next month are somewhat erratic and unpredictable. They may stay together in large herds, drifting slowly northward, or they may scatter in small groups over the nearly three hundred thousand square miles of their summer range.

July is the month of new grass and sedges. Bumblebees dart busily over acres covered with blossoms. Spiders lie in wait at the entrance to their silken lairs, ready to pounce on any passing insect. It is also the month when the tundra air is filled with the shrill whine of myriad mosquitoes, following man and beast as a voracious, humming halo. Despite the assertions of northerners that their mosquitoes are much bigger than elsewhere (a popular saying is "twelve can drain a rabbit"), this is not true. They are only infinitely more numerous.

A raven glides over the Barrens. This scavenger is by far the earliest bird to nest in the Arctic. It nests on cliffs in late March or early April. Very often the nest is later taken over by gyrfalcons.

The nineteenth century German zoologist Alfred E. Brehm, travelling across the tundra in Russia's Arctic encountered there the winged hordes and wrote with the poetic venom of a much-stung man: "They form swarms which look like thick black smoke; they surround, as with a fog, every creature which ventures into their domain; they fill the air in such numbers that one hardly dares to breathe; they baffle every attempt to drive them off; they transform the strongest man into an irresolute weakling, his anger into fear, his curses into groans."[6]

Seton, who had a passion for statistics, pulled his glove off, held out his unprotected hand, then counted the mosquitoes that alighted on it. His usual score was 100 to 125 mosquitoes in five seconds.

They would drive me nearly out of my mind when I photographed flowers. The best days for this were when it was calm and sunny, precisely

(Left) In July flowers carpet square miles of the Barrens and flies flit busily from bloom to bloom. (Right) A tundra spider waits at the entrance to its silken lair. (Below) Incredibly numerous, the Lapland longspurs seem to have their cup-shaped nests in every tenth tussock.

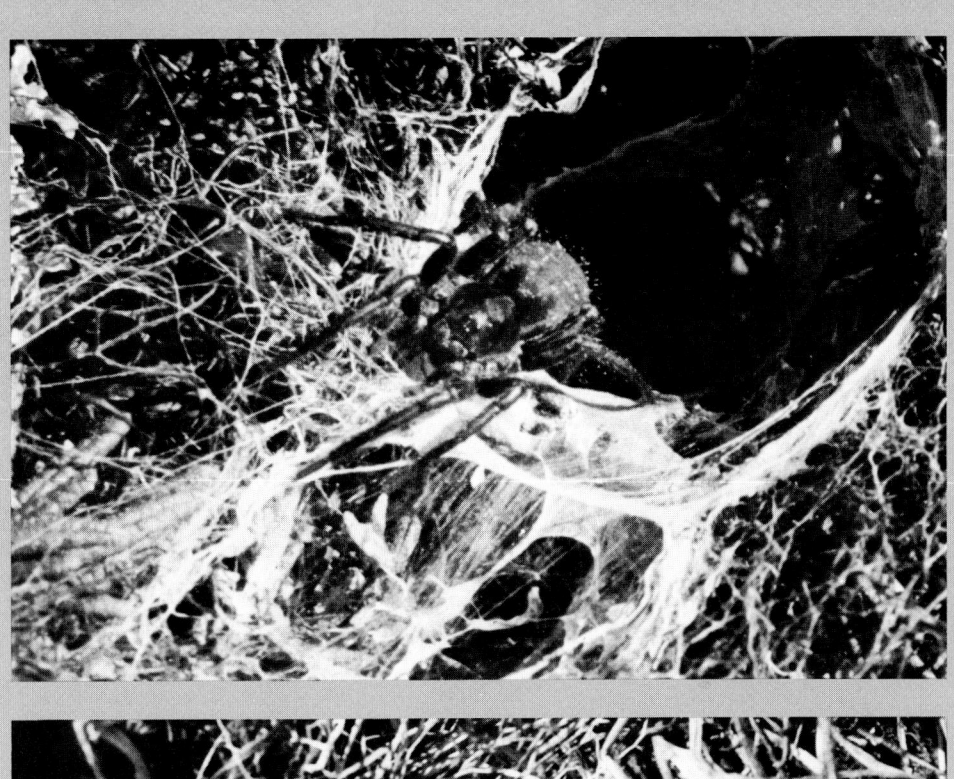

Caribou cows and calves migrating across the Barrens.

the conditions mosquitoes love for launching a mass attack. Not that they could do much. I was heavily dressed, wore leather gauntlets and a great black veil, and this they could not penetrate. But they settled on the stiff gauze in a nearly solid layer, and it's a difficult task to focus a camera when one's vision is screened by a mass of mosquitoes. The male mosquitoes just swarm about, like glittering wisps of smoke in the sunset, but the females are out for blood and they make life miserable for beast and man.

The caribou suffer intensely from this midsummer insect plague and this may account, in part, for some of their erratic wanderings. To avoid mosquitoes and black flies they like to walk against the wind, or to seek out wind-swept hills. On calm days, they stand in a circle, as tightly pressed together as possible, their heads lowered to the ground. I once saw such a herd from a plane, and it looked most peculiar – as if they were holding a prayer meeting.

While mosquitoes and black flies may harass the caribou, the insect they really fear is the warble fly, a rather attractive hirsute animal looking like a bumblebee that's been on a diet. It has a loud distinctive buzz, and caribou react violently to this noise. They break into a frantic gallop, twitch and shake vehemently, and race off again in wild desperation. Apparently these evasive manoeuvres do not help much, because nearly all adult caribou are infested with warble fly larvae. They tunnel along under the skin of the caribou, bore their way through it when full-grown, drop to the ground and pupate, leaving the skin so full of holes and scars that it looks as if it has been peppered with buckshot. The Eskimos once regarded the inch-long, oval, cream-colored larvae as a gustatory treat, and Stefansson, ready to try anything, ate some and reported they tasted like gooseberries.

Though caribou really fear the warble fly, they seem more relaxed about their most deadly enemy (after man) – the wolf. If a wolf attacks a herd, those animals singled out for pursuit flee, while the others show little alarm. Caribou seem able to tell when a wolf is merely out for a walk and not hunting, and then they nearly ignore him.

When man makes a mess, his first reaction is usually to look for someone to blame it on. Thus, when it was belatedly realized that caribou numbers had declined drastically, it was only natural to blame it all on the big bad wolf. They were poisoned on a large scale, more than ten thousand just in the years between 1953 and 1959 throughout the caribou range.

Yet there is considerable evidence that wolves take primarily sick and weak animals and may, in fact, contribute to the health and vitality of the herds. While following caribou by aircraft I noted that one rarely sees wolves in the immediate vicinity of the herds. Usually they lag a few miles behind, presumably to pick off stragglers. It seems doubtful whether the

The newborn caribou calf seems to be all gangly legs. In two days it will be able to outrun a man.

The small, non-migratory, light-colored Peary caribou of the high Arctic islands.

wholesale extermination of wolves would help the caribou increase at a much faster rate, and it could expose the herds to devastation by diseases which might have been checked, had the sick animals been eliminated by wolves. Extermination would also deprive the tundra of one of its most beautiful animals, a predator that has lived in balance with his prey through many millennia.

Wolves, unfortunately, have always had a bad press, starting with the fairy tales, and in the popular mind they remain horrid, ravening monsters that gobble up grandmothers and ravish Little Red Riding Hoods. Quite intelligent people have asked me whether I carry a gun in the North, and when I say "no," they exclaim: "But aren't you afraid of wolves?"

In fact it is most difficult even to see a wolf. They are extremely wary and flee long before a human has even spotted them. Only once was I lucky enough to see wolves close. It was a drizzly day on the Barrens. I lay behind a rock, watching snow buntings, hoping to find their nest. Suddenly I saw something move in the distance. The binoculars showed it to be a big grey wolf, with black, mask-like marks on its face. It trotted leisurely across the meadow, lay down a hundred yards away, scratched itself with its hind foot like a dog, curled up and went to sleep. After half an hour, the wolf got up, stretched itself, yawned, and crossed the meadow. It ran to the top of a large rock outcrop, stood still for a minute in superb silhouette, and disappeared on the other side.

I thought it had gone, and walked to the rock. Just as I got to it, I saw another wolf, this one nearly pure white, trot across an esker bordering the meadow. I lay down and watched. The white wolf came straight towards my rock, and when he was fifty yards away – the grey one emerged from the other side of the rock outcrop, where she had lain only a few yards from me.

The white male advanced purposefully, the grey she-wolf went a bit hesitatingly to meet him. They met and sniffed, tails wagging. Suddenly the grey wolf rushed, the male fended her off with a stiff shoulder. They played tag, chased each other, one bounded against the other in feigned fury and they rolled over and over, two beautiful, playful animals, having a glorious time. They flushed a ptarmigan. It landed on a rock, burping excitedly. The male "stalked" it, not stealthily, but just for fun, walking towards the agitated little cock in an exaggeratedly stiff-legged manner. They were only thirty yards from me. Once I thought the female had spotted me. She stared intently (in the binoculars she looked four feet away), the black facial stripes giving her an odd quizzical expression. But she decided everything was fine, joined the male, and they loped off together, jostling each other from time to time, just for the heck of it, free and wild and beautiful.

One summer I stayed with Samson Koeenagnak, one of the last Eskimos to live far out on the Barrens, at Aberdeen Lake. We shifted camp, and at

(Above and right) A snowshoe hare, white in late fall, is conspicuous among the dark tundra boulders. (Top) An arctic hare.

the new site there was a lovely, nearly white bitch, three-quarters wolf, the daughter of a female husky, herself half wolf, who had mated with a free wolf. She had run away and had supported herself quite well for many months, but now she returned to our camp. She seemed friendly enough, but the Eskimos did not trust her, nor did she ever come close enough to be touched.

Before we recrossed the lake Koeenagnak threw her a chunk of meat, slipped a noose over her head as she picked it up, strangled her into submission, tied a thong tightly around her muzzle and strung her up against the bow-thwart of our canoe. The rest of the boat was piled with the family's belongings, plus Koeenagnak, his wife, their children and nine hefty huskies. I sat in the bow. I felt sorry for the wolf-bitch, hanging there and asked whether I could release her. Koeenagnak looked dubious, finally lifted his eyebrows in assent. But she was still unhappy. It was hot and all the other dogs were perspiring with lolling tongues. Gently, talking to her all the time, I unwound the thong from her muzzle. She growled deep in her throat and watched me with slanted, pale blue eyes. But she didn't snap, and settled, quietly breathing, at my feet, and after a while I forgot about her. Later some terns flew over our boat and hovered above us. Absentmindedly I reached down for a camera, touched the wolf, and iron jaws clamped onto my wrist, jaws strong enough to crunch a caribou bone with one bite. I froze with fear, but the wolf-bitch did not bite. She held me a moment and then let go. Her long fangs had not even penetrated the skin. It had merely been a warning not to touch her. She was a gentle wolf.

Summer is brief on the Barrens. In a few short weeks in June and July all nature's pent-up vitality seems to burst forth in a passionate renewal of life. Flowers carpet the tundra for miles, but in August the dainty petals turn pale, shrivel and drop. Among the millions of emerging young animals there is a great urgency to grow, to complete life's cycle – from birth to independence from parental care – within summer's brief span.

The caribou are resplendent now in their new coats of a rich clove brown, the legs nearly black and set off with strips of pure white fur just above the hoofs. The underside and flanks are white, and bulls grow a flowing white throat mane. They are bunching again into large herds and slowly begin to drift south. Until recently it was assumed that caribou make one big round trip: up to the tundra in spring and summer and down to the forest belt for the winter. Now, although the reasons for it are not quite clear, it has been established that the migration is really one of double flux and reflux: into the tundra in spring, back to the forest (or at least its edge) in August, and northward again in September, though not as far as during the spring migration. The bulls at this time may form separate herds but rejoin the main herds at the end of that month. At the time of the rut, in

the middle of October, the caribou are still on the Barrens, but already are moving purposefully southward, to scatter into groups throughout their roughly three-hundred-thousand-square-mile winter range in the taiga, after a total migration of from two to four thousand miles.

I was at Koeenagnak's camp, at the narrows between Aberdeen Lake and Beverly Lake when the caribou marched southward in late July. Here caribou have crossed, and Eskimos have waited for them, since time immemorial. On this major migration route the sharp-edged hoofs of untold generations of caribou had worn deep parallel paths into the tundra soil. The caribou flowed past our camp for three days and four nights, sometimes in groups, sometimes in single file, paying little attention to us and occasionally coming to within thirty yards of our tent and the line of tethered dogs.

Later, from Baker Lake, Jerry Parker and I followed a large herd by plane. During the day the animals were usually scattered over many square miles. Towards evening some began to move southward, still feeding, but drifting in a definite direction at the same time. Gradually others joined them, they fed less and moved more, the haphazard drift became a purposeful march and the many-mile-long mass seemed to glide across the tundra, skirting lakes and bunching into tight brown clusters at rivers and narrows. Some animals began to swim, and soon a long string of caribou crossed the Styx-dark waters; and on they marched, ever on, over "the vast and dreary plains" as Tyrrell called the Barrens. The sun was setting. The land lay hushed and somber, its myriad glittering lakes reflecting the copper sky. Over the dark earth snaked the vast throng of migrating caribou, a golden ribbon of life in the sun's slanting rays.

Early September brings the first frost. Delicately-veined ice glazes the smaller ponds. The siksiks, fatter than ever, retire to their burrows for their long sleep. Near the lake shores, plovers and sandpipers trip busily through the muck, leaving a filigree of slender three-toed tracks. Twittering flocks of small birds flit over the tundra and high overhead skeins of geese wing purposefully southward.

Fall's first snow brings the stillness of winter to the Barrens—this vast and beautiful land which knew in the few intense months of spring and summer the fullness of life.

(Top left and right) A semipalmated plover in a lush tundra meadow and on its nest next to a stone. Spangled in bronze and white, the golden plover (right) is one of the most beautiful birds of the Barrens. It lures enemies away from its nest (below) by pretending to have a broken wing. (Overleaf) Canada geese against the evening sky.

A spiky pile of antlers near an Eskimo camp. In the settlements, Eskimos make carvings of them. Long ago Eskimos waited in this stone ambush (facing page) for the arrival of the caribou herds.

"Ager-fauk that is milke white"

Much of the Barrens, that immense tundra region west of Hudson Bay, and between the treeline and the Arctic Ocean, is low, rolling country. But south of Baker Lake the great Kazan River has gnawed its way through a rocky barrier and hurtles its whirling waters down a cascade and on through a two-mile, cliff-flanked gorge.

We landed on a lake above the falls to inspect the carcass of a caribou killed by wolves, and then walked to the gorge. As we came near, a pair of rough-legged hawks circled upward, their melancholy mewing cries strangely in harmony with this vast and lonely land. As I passed near the cliff's edge, a peregrine falcon flew up with a startled cry. Its nest, a mere scrape on a grassy ledge, was just a few feet below the top of the cliff. Two speckled eggs lay in it. A quarter of a mile further, a second falcon flew up. Another peregrine, I thought at first, but it was larger, and lighter in color, and I realized with a start of pleasure that this was a gyrfalcon.

The large nest, probably built originally by a rough-legged hawk or a raven, was on an inaccessible shelf, protected by a slanting rock overhang. Not far from it was a crack in the cliff. Squeezing downward through it, I emerged some ten feet below on a narrow ledge. From there I could see the nest and the three nearly fully fledged eyases, as young falcon are called.

It was an awkward place. The ledge was not even a foot wide, and there was nothing to hang onto. The balancing act wasn't made any easier by the angry female falcon. Unlike the peregrine, who spirals up to gain altitude for his bolt-like, 180-mile-an-hour stoop, the female gyrfalcon flew upwards in a straight line with fast powerful wingbeats until she was but a speck in the sky. There she veered and came racing back, wings half-closed, like a feathered bullet, and sheered past me with a shriek of fury, trying to scare me off my precarious perch. Two hundred feet below, the tumultuous river rushed along in swirling eddies. Jerry Parker, full of

(Preceding pages) A white gyrfalcon at its nest in the high Arctic. (Below and right) A gyrfalcon on its nest on a ledge protected by a rock overhang, high above the turbulent Kazan River rapids. Two young gyrfalcons (bottom right) glare at an intruder.

kindly concern, looked down the cliff and said: "For heaven's sake, Fred, don't fall in! It would spoil my whole summer!"

Two of the young falcons lay pressed against the rock wall. The third stood on the nest, and fixed me with a haughty stare. I looked at it with reverence. This was the legendary gyr, bird of kings, and king of birds.

Falconry is an ancient sport. In Assyria it was popular as early as the time of King Sargon II in the seventh century B.C.; and at one time or another nearly every bird of prey has been pressed into man's service (except the owl and the vulture), from the mighty eagle, flown by Kirghiz horsemen galloping across the steppes of Asia in pursuit of wolves and foxes, to the dainty merlin, perched on the gloved hand of a medieval lady draped in flowing skirts and riding sidesaddle after blackbirds and robins.

From the Middle East, falconry spread to Europe in the ninth century. It became an obsession with medieval man, and the birds themselves the status symbols of a hierarchical society. The lower aristocracy had to content themselves with goshawks and sparrowhawks, and a mere burgher was lucky if he was allowed to fly a kite. The merlin was the ladies' falcon, and the hobby was flown by young men of noble birth. Counts and earls could hunt with the saker, the lanner or the peregrine. But the gyr, largest of all falcons, was the bird of emperors, and kings, and dukes, and princes of the church.

"A ger-fauk that is milke white," says the fourteenth century English romance *Guy of Warwick*, was a symbol of position, prestige and wealth. Each bird was worth many times its weight in gold. When the son of the Duke of Burgundy was captured in 1395 by the Ottoman Sultan Bayezid, his father offered two hundred thousand gold ducats in ransom. The sultan instead demanded (and got) twelve gyrfalcon.

Falcons and falconry played a large role at most courts. Under King Francis I of France (1515-1547) the Grand Falcon was in charge of fifty "gentlemen falconers" and about seventy "falconers" who, one gathers, were not gentlemen and presumably did most of the work. When the king moved court, the whole feathered retinue followed – falcons and falconers. And in China, Marco Polo reported, when the Emperor Kublai Khan went hunting he was attended by ten thousand falconers, carrying gyrfalcons, peregrine falcons and sakers.

The *haute volerie* was not only the passion of lay nobles. Clerics, too, loved it and a special papal edict permitted them, on hunting days, to say mass booted and spurred. Their favorite falcon was the great white gyr, and these *falcones alba* were sent as tribute to the papal see. In fact the earliest known falcon-catching licence was issued in 1194 by Pope Collestin III to Archbishop Erik of Nidaros, the present-day Trondheim, in Norway. Severe laws protected the valuable birds. From the man who stole

A rough-legged hawk soaring above the northern tundra. The arctic summer is short and the hawk has laid a full clutch of eggs even before the ice has gone out of the river below.

a falcon, medieval law exacted half a pound of flesh, to be cut from his chest and fed to the falcon.

Unlike the peregrine, common in Europe and easily caught, gyrfalcon had to be imported from the lands of Ultima Thule: first from northern Norway, and later from Iceland and even Greenland.

This great arctic falcon occurs in three color phases. In the dark phase, with upper parts a fairly uniform brown, it inhabits the southernmost region of the gyrfalcon range: southern Greenland, Labrador, the Barren Grounds, and parts of Alaska and Siberia. Audubon found a gyrfalcon nest in 1833 near Bradore, close to the Strait of Belle Isle, across from northern Newfoundland; but this was exceptionally far south.

Falcons of the grey phase, whose crown and nape are mixed with white, occupy territory further to the north. In the high Arctic, breeding as far north as Peary Land on Greenland, within about five hundred miles of the Pole, live the pure white falcons, their snowy plumage delicately marked with dark bars and spots. In winter, when darkness descends upon the far north and the sea birds leave and even the ptarmigan migrate, the white falcons fly south, some to southern Greenland, others across the sea to Iceland.

It was these white gyrfalcon, with their nearly four-foot wingspread, their tremendous speed and power and their deadly lightning-like stoop, that were most highly prized. Once the Vikings settled in Greenland, gyrfalcon became one of their most valuable exports, and they are the only birds mentioned in one of the oldest records about Greenland "The King's

A young peregrine falcon.

Mirror." When Iceland came under Danish rule, the Danes sent over falcon catchers, and the island paid part of its tithes in gyrfalcon for the royal mews.

Indians and Eskimos, on the whole, preferred a falcon in the pot to one on the hand. Young gyrs and peregrines were taken from nests in Greenland and eaten. The eighteenth century explorer Samuel Hearne, reported gyrfalcon were the only "hawks . . . [that] brave the intense cold of long Winters to the North of Churchill River." Because the gyrfalcon preyed heavily on ptarmigan, the resident Hudson's Bay Company governors at Churchill "generally give a reward of a quart of brandy for each of their heads. Their flesh is always eaten by the Indians and sometimes by the English; but it is always black, hard and tough, and sometimes has a bitter taste. The Indians are fond of taming those birds and frequently keep them the whole summer; but as the Winter approaches they generally take flight and provide for themselves."[1]

The gyrfalcon breeds at least a month earlier than the peregrine, in April in its southern range, in May in the high Arctic. Since snow still covers land and cliff ledges, they nearly invariably nest on a shelf kept snow-free by an overhang. Their own efforts at nest building are as rudimentary as those of the peregrines, consisting of a mere shallow scrape. They nearly always take over large nests built in a previous year by either a raven or a rough-legged hawk. Like the peregrine, gyrfalcon will return to the same eyrie year after year. In Britain, peregrines are known to have occupied favorite nesting sites nearly without interruption since Elizabethan times.

In Greenland, Iceland and Norway, gyrfalcon prefer to have their nests in the vicinity of sea bird colonies, where they are assured of a steady and ample food supply. There they usually lay three or four eggs and raise their young without much trouble. The falcons of the Barrens, feeding largely on ptarmigan and lemming, both cyclic animals, may lay an extra large clutch of eggs and raise all eyases successfully in years when these animals are plentiful. But when their cycle is low, the gyrs may not even bother to nest, or may only raise one or two young.

The gyrfalcon, whose name comes from the Old High German *giri* – "greedy," live up to this appellation. Observers have estimated one pair and its young may polish off two hundred ptarmigan in one short summer, mainly the white and conspicuous cock ptarmigan.

The peregrine lives nearly exclusively on birds. It spirals upward, then stoops, wings folded, like a feathered bullet, killing small birds on impact or squeezing the life out of them with long, curved talons. Larger birds are sometimes approached from the side, or the peregrine may fly underneath them, turn onto his back and pluck his prey out of the air with an extended foot.

While the female gyrfalcon broods, the tercel hunts for her. His method of hunting varies, according to prey, from lightning stoop, to stealthy stalk. Sea birds are taken on the wing, and once a falcon has singled out a victim, only an instantaneous dive into water can save it. On the tundra, where prey is more terrestrial, the falcon usually flies low, then dives to snatch up an unwary groundsquirrel or ptarmigan.

The young hatch after twenty-eight or twenty-nine days and from then on both parents are busy lugging in food to still the appetite of their voracious brood. The tercel usually passes his catch to the female, who carries it to the eyrie, tears off bits of meat and feeds it to the eyases.

It is usually easy to find the nests of peregrines, gyrs or rough-legged hawks. The cliff around the nest is splashed with a bright-orange nitrophilous lichen. One can spot them even from low-flying aircraft. Usually each bird of prey has a "territory," and consequently nesting pairs are widely spaced. But in the Barrens, where suitable nesting cliffs are at a premium, I have often seen the nests of rough-legged hawks and peregrines quite close together, which must be a nuisance for the poor hawks, because the peregrines, instead of acting neighborly, insist on harrying the hawks whenever they fly up. At the Kazan River rapids, I found three nests in half an hour on just one cliff-side of the river.

I saw an even more impressive example of the attraction of cliffs further south, near Cape Henrietta Maria, where Hudson Bay and James Bay meet. This is a game-rich but absolutely flat tundra area. There isn't a good cliff within a hundred miles. Fifteen miles from the coast is an abandoned DEW-Line station. The houses still stand, the machinery, and three giant concave radar antennae. Near their tops, on the joints of the steel-beam supports, I found five nests. It was late fall, and the nests were abandoned but all showed signs of recent occupation. They looked as if they had been built by rough-legged hawks or ravens but, since so many pairs of the same species would presumably not nest so near each other, it seemed likely that falcons, too, had occupied these man-made cliffs during the preceding summer.

The gyrfalcon from the high Arctic migrate southward in fall to spend winter, roughly in the areas occupied by their duskier cousins. A few wander into southern Canada and occasionally into the northern United States. Only twice in the last century have gyrfalcons been seen in Britain. While the gyr prefers the north at all seasons, peregrines fly far to the south in winter, into the United States and some even to South America.

In Greenland, migrating gyrfalcon used to be shot "for sport." In 1937, white trappers at Scoresby Sound counted three hundred gyrfalcon in eighteen days, and gunned down two hundred and fifty of them. Since then, such massive slaughter of these beautiful birds has ceased, and throughout the rest of their range they are rarely hunted. In Iceland

(Below) Downy but sharp-billed, a peregrine falcon eyas on the ledge of an arctic cliff. (Left) A young peregrine falcon on Southampton Island.

gyrfalcon are completely protected – rather fitting since this bird is the country's national emblem. It is estimated that there are about two hundred nesting pairs of gyrfalcon in Iceland, between two hundred and three hundred pairs in Alaska and about one thousand pairs in Greenland. The total world population of gyrfalcon is estimated at about five thousand individuals, compared to an estimated thirty thousand peregrine falcons in the northern hemisphere.

Although they are no longer extensively hunted, a new and possibly graver danger now threatens these magnificent falcons. Pesticides are used extensively and small birds pick them up with their food. In the falcons who live on these birds, the poison becomes concentrated. Concentrations of DDT can render falcons sterile, or cause them to lay thin-shelled eggs which do not hatch, and the survival of these species becomes threatened.

Already the peregrine has become rare in large areas of the United States and Europe where, only a few decades ago, it was widespread and numerous. Even the gyrfalcon is not safe in its remote arctic realm. Significant amounts of pesticide have been found in their eggs and, although they live beyond the range of actual use of these poisons, the birds on which they prey do not. Thus the great white falcon, once the prized possession of popes and princes, may now only be safe from man's poison in the northernmost reaches of its frigid kingdom.

Birds of the Sun

Our base camp on Spitsbergen was on an extensive raised beach. Near the camp were two little fresh-water ponds. From one we fetched water, in the other the hardier, or foolhardier, members of the expedition went for early-morning swims. To reach either pond we had to pass through a breeding colony of arctic terns and these graceful little birds bitterly resented our intrusion. They screeched their rage and dive-bombed us with precision, hitting our heads with stiletto-beaks that drew blood. In addition to inflicting this avian acupuncture, they disgorged on us the remains of their last meal until, bleeding and spattered, we rushed back to camp.

    The terns arrived in mid-June and, since the season is short, started courting at once. A male flew off to the coast, hovered on rapidly beating wings above the water, head bent down. Suddenly he closed his wings, dropped like an arrow, and emerged an instant later with a glistening little fish. Crying in triumph – how they can make such a racket without dropping the fish, is something that always mystifies me – the male flew

Superb fliers, arctic terns nest in the far north and migrate as far south as the Antarctic in winter.

(Left) Elegant wings held high, an arctic tern settles on its nest. (Right) Screaming in rage, terns whirl around a trespasser upon their territory.

back to his chosen female. He strutted around her on tiny coral-red feet, holding out the fish. Then another male flew up, also with a fishy offering. The female, more practical than romantic, accepted both fish with impartial appetite, and let the angry males fight it out amongst themselves. The winner then rushed back to the coast for another fish to offer to the female. Once she accepted it, that seemed as good as a "yes" in church, and she would firmly reject all further fish-bearing suitors.

Their nests were rudimentary, just a shallow hollow in the sand or gravel. In each two eggs were deposited, faithfully brooded and valiantly defended. When the downy chicks emerged both parents were busy shuttling back and forth between sea and nest, stuffing those little gluttons. Once I came upon a pot-bellied chick, half a fish sticking from its bill. There just wasn't room inside for all of it. The little tern sat back on its rear-end, swallowed hard every few minutes and the fish slid in a bit further. Finally only its tail protruded. The chick made one last great gasping gulp and that, too, disappeared.

The little terns were in a hurry to grow and gain strength, for soon they would have to make the longest migration of any bird. From their birthplace in the far north, only eight hundred miles from the Pole, they would fly along the coast of Europe and of Africa to its southernmost tip, and from there some might continue further south, across the gale-whipped

(Left) Glaucous gulls. (Right) The ivory gull is much smaller than the glaucous and is all white.

Flying down to pick food off the sea, a jaeger meets its mirrored image in the calm water.

waters of the "roaring forties" and the "shrieking fifties" right down to the Antarctic, spanning the antipodes in their eleven-thousand-mile flight. When they were born the sun never set, but circled day and night above them in the arctic sky. When they left, the sun did not yet dip below the horizon and they would follow it to the other side of the world, for another summer with long daylight hours. Thus the arctic terns not only have the longest migration, they also enjoy more daylight than any other bird.

Arctic terns hatched in the Canadian Arctic make an even longer flight. Some first see the sun at the extreme north tip of Ellesmere Island, a mere five hundred miles from the North Pole. In fall they fly south along the west coast of Greenland, cross the Atlantic to Europe and join their cousins from the European Arctic in their flight along the African coast. While most fly south, some fly west from Africa, cross the Atlantic a second time, and continue towards the Antarctic along the west coast of South America. By the time they reach the Arctic again, next summer, they will have flown more than twenty-two thousand miles. In addition to these long-distance trips, terns fly a great deal at their summer and winter homes. One banded tern is known to have lived to an age of twenty-seven years, and it is pretty safe to assume that this small, graceful bird had flown close to a million miles in its life.

With due respect for such accomplishments, their daily attacks on our haulers of water were a nuisance until we found a simple method to keep them off. If we held a stick above our heads the terns would not come near us, presumably for fear of impaling themselves. They just hovered above us and shrieked in anger. Oddly enough, the terns at the colonies I visited in the Canadian Arctic, seemed less aggressive. They dive-bombed, but seldom actually pecked me.

If terns can harass humans who trespass on their territory, they in turn are harassed by the master marauders of the Arctic, the jaegers. These graceful, gull-like birds, with hawk-like habits, catch some of their own prey. Especially the great pomarine jaeger (or "skua" as the jaegers are called in Britain) will methodically hunt lemmings in years when these little rodents are plentiful, killing them with its strong hooked beak. All the jaegers are inveterate nest robbers, filching eggs and killings young birds. By inclination, they are pirates, content to let others work and then to plunder them.

Gulls and terns fly well, but jaegers can fly circles around them. Their method of stealing is simple. They sit on a rock and wait. When a gull or tern returns from the sea, the crop or bill full of fish, the jaegers throw themselves aloft and start to chase their victim with loud, raucous cries. They mob him. They stoop from above, attack from the side, sail up from underneath and make such a nuisance of themselves that the gull or tern disgorges or drops its food. Instantly the jaegers swoop down and pluck

The gregarious kittiwakes nest on arctic cliffs. They spend the winters at sea.

181

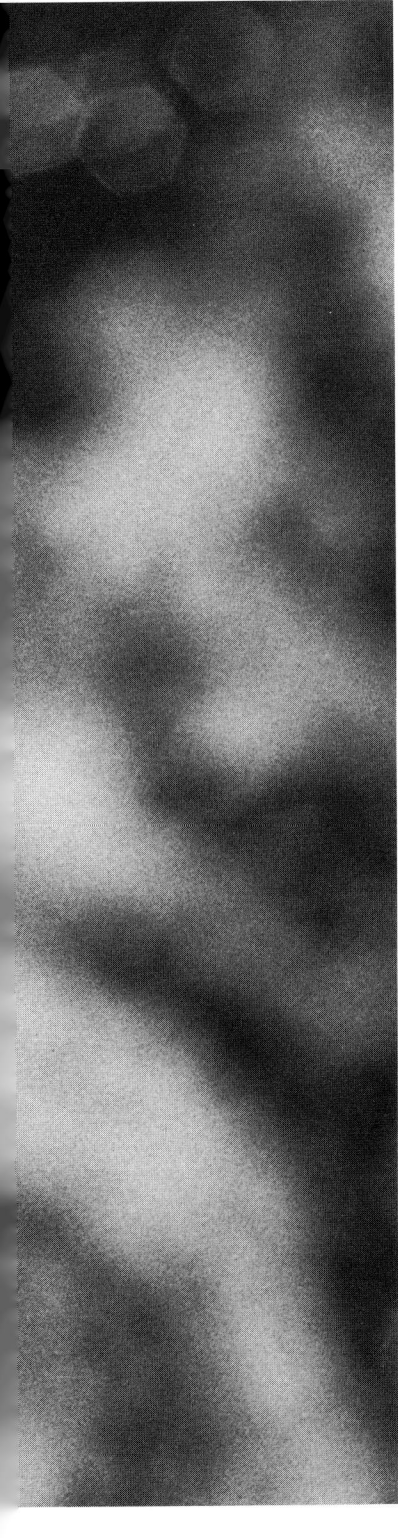

Every ledge and cornice on a sheer island cliff is used by kittiwakes to support their nests.

Fulmars can be distinguished from gulls in flight by their much stiffer wingbeats. They occur in dark and light color phases.

the tidbits out of the air. While their victim flies back to sea, the jaegers return to their roost, to wait for someone else to come by whom they can rob.

If anything, they are even more aggressive towards human intruders than are terns. I found a nest of the parasitic jaeger on a tundra tussock, amidst a rather marshy area. Both parents accompanied my progress towards the nest by attacking me with ever increasing fury, hitting me with bills and hard wings.

While the jaegers resent humans near their nest, they showed no qualms about visiting us at our camp on the tundra to filch caribou meat. At first they were only a few, but the good news must have spread and within a week we were host to a noisy horde of some thirty jaegers. We did not grudge them the food but we did resent the racket. They shrieked, and wailed, and mewed, and quarreled endlessly about every bit of gut and meat.

Like the terns, the jaegers are long-distance migrants. Some of our noisy friends would leave the Arctic in August, fly out to sea, following the coasts of North and South America, all the way to the Falkland Islands or Tierra del Fuego, and return again early the next year to their northern home after a round-trip of sixteen thousand miles or more.

Usually the jaegers pursue, pester and plunder only the smaller gulls, particularly kittiwakes, but on Coats Island they did not hesitate to harass the great glaucous gulls.

Because the glaucous gull is so big and bossy, whalers used to call it "burgomaster," the old Germanic term for borough master, or mayor. It is a beautiful gull, pure white with a pale-grey mantle. On Coats Island they had a breeding colony on a large, sheer cliff bordering a beautiful lush green valley. The cliff was covered with vivid splashes of nitrophilous lichen, and against this bright-orange tapestry soared the great white gulls to and from their nests on the ledges of the cliff.

They come early in spring to the north and hunt for fish or crustaceans in the leads among the ice. Later, in addition to fishing and assiduous beachcombing, they raid the nests of other birds. On Coats Island glaucous gulls also had their nests on the great murre cliffs, just above the main colonies, a very handy arrangement, tantamount to living in a well-stocked larder.

After a long walk and climb, I once approached the gull cliff overlooking the valley, from above. The nests were on inaccessible ledges below me, quite large structures of grass, seaweed, lichen, moss and small sticks. Three olive-brown eggs, splotched and splattered with dark brown, were in the nests I could see. I had hoped to photograph the gulls from above on their nests, but a flying gull gave the alarm, and most of the brooding birds soared into the air. Those that remained seemed to have a shrewd idea whence danger came, because when I poked my head over the

Fulmars squabble over bits of blubber. The nasal tube along the top of the bill is clearly visible.

edge of the cliff they were all looking up, spotted me instantly and followed the others, soaring beautifully in the head-wind, the wings now widely-spread, now half-folded, as if they were moulding themselves to the air currents.

In former days, Eskimos used to raid the Coats Island gulleries and occasionally still they take eggs from the cliffs at Duke of York Bay on northern Southampton Island. Now the eggs are usually eaten within a few days. In the past, though, Eskimos stored them in ice- or snow-covered, shaded hollows, where they stayed reasonably fresh all summer.

Superb fliers that they are, the aerobatics of the glaucous gulls are as nothing compared to those of the fulmar, the master glider of the Arctic. Fulmars are related to the shearwaters and, a bit more remotely, to the albatrosses. They don't look elegant. The wings are long and narrow, the body bulky and nearly neckless, the bill is thick and stubby, surmounted by a conspicuous nasal tube which is divided by a septum into two nostrils. Nor is their flight particularly impressive at first glance: a series of rapid wingbeats followed by a glide.

To see them at their masterful best, there should be a storm, huge waves should race each other in foaming succession, separated by deep troughs. That's fulmar weather. They glide low over the tumultuous tops on stiff wings, dive down into a trough, and soar upward on the vertical air currents formed on the windward side of each wave trough. They bank sharply, swoop down again and rise up, skim across the crest seemingly inches from the grasping water and slide down effortlessly its far side, to surge upwards again.

Fulmars are oceanic birds. They spend the winter at sea, soaring over the waves, often on the great fishing banks where they follow trawlers in hope of scraps. Newfoundlanders call them "noddies"; to the whalers they were "mollies"; and fishermen also call them "oil birds," since a captured fulmar will spew out a good tablespoonful of vile smelling oily liquid. For this reason they are also called "stinkers"!

They eat fish, cephalopods, various crustaceans and those shrimp-like euphausiids known to fishermen as "krill." They are crazy about fat. In the days of arctic whaling they used to surround the whalers in large, squabbling groups as whales were flensed in the water alongship.

When we caught narwhal in Koluktoo Bay it didn't take the fulmars long to come and claim their share of fat. When they fly and glide they are silent, but throw them a lump of fat and they will squawk like a bevy of bilious ducks. The way they gorged themselves on blubber – rip and gulp, rip and gulp – one would have thought they hadn't eaten in years, and when they were finally full-up, they could barely fly.

Fulmars are voracious scavengers and often follow ships in arctic seas.

Fulmars come early in spring to the Arctic. I was travelling with some Eskimos by dogteam across Jones Sound, between Ellesmere Island and Devon Island, in early May. It was a cold, windy, white-outish sort of day. We had sledged across the frozen sea for days without seeing a living thing. Suddenly out of that greyish pall sailed a flock of fulmars, and banked silently past us on ash-grey wings.

Fulmars were the "northernmost" bird seen by Nansen's *Fram* expedition, at more than 85 degrees north, less than five degrees from the Pole. The following year, 1896, as the ice-held ship drifted towards Spitsbergen, fulmars became even more numerous, and more than seventy were shot by the crew as dog food.

Their breeding grounds are on sheer cliffs in the high Arctic and the colonies can be immense: more than a quarter of a million fulmars are estimated to inhabit Cape Searle on southeastern Baffin Island. The fulmars arrive early in spring and with good reason. For an arctic bird, it takes an inordinately long time, more than one hundred days, from the time the single egg is laid until the young is fledged. The solitary egg, a lusterless white, is incubated by both parents in turn, for nearly two months. What finally emerges makes the prolonged effort, to human eyes, hardly worth while: a little lump of life, wrapped in whitish down.

Most fulmar colonies are on cliffs facing the sea. Some, though, are far inland. When we crossed Spitsbergen's great icecap, we were struck by the absence of life. For days on end the only beings to keep us company were the beautiful little ivory gulls. Pure white, with little black feet and a black bill, they are the daintiest and, to me, the most beautiful of all gulls. The idea that they are like incarnations of celestial beings is somewhat shattered by the realization that their main reason for following us (they also follow dogteams in the Arctic) was to eat our excrement.

In the center of the icecap large *nunataks* (ice-free mountains) soared up, and on their cliffs, amid this immense white carapace of ice, fulmars sat by the thousands on their eggs, and in the snow below the cliffs little tracks showed that arctic foxes knew all about this potential source of food.

The single chick is stuffed by both parents, who squirt the squab full of a regurgitated amber-colored oil. On this oleaginous diet the little fulmar expands nearly visibly and in three weeks has put on such a layer of fat that it weighs more than its parents. Having done their duty, the adults leave, and the fat little fulmars sit on their ledges for another three weeks, absorbing some of that surplus fat. Even then they are so corpulent they cannot fly but merely glide down to the sea, where their fast continues until they have slimmed down enough to be able to lift themselves off the water. Then off they fly to the open sea, soaring over storm-tossed waves, gliding on stiff wings from crest to trough and trough to crest – supreme masters of the air currents.

The Downy Ducks

South of the Lofoten Islands, off the coast of arctic Norway, lies the tiny and isolated archipelago of Röst. It consists of 365 holms (small, grass-covered islands) and 1,800 skerries (rugged, bare, rocky islets), many flooded at high tide, favorite roosts of shags, puffins, murres and gulls.

The steamer from Lofoten deposited me at Röst in the evening, and I found a room at the "fishermen's home," now nearly deserted since the cod fishing season had ended. It was warm and I left the window open, and throughout the spring night the melodious crooning of love-stricken eider drakes drifted into my room, *ah-ooo, ah-ooo, ah-ooo*.

There were eider nests everywhere: under old boats, left as shelters for the ducks; in broken barrels; in little shelters built of flat stones; and underneath the raised houses. One duck scared the daylights out of me while I was using an outdoor privy. She had her nest in a dark corner of this building, where I hadn't noticed her, and burst into my meditations with a loud squawk.

One day I visited Thoralf who, like many of Röst's individualistic fishermen, had an island to himself. I tied the boat to a wharf post, climbed the ladder and nearly fell backwards into the water when an angry duck on the wharf, near the top of the ladder, hissed into my face. The door to Thoralf's house was open. "Come in," he said hospitably. "Have some coffee." Norwegian fishermen's coffee is a potent brew. It sits all day on the stove and simmers until it's strong enough to make a heart of stone palpitate. We drank our coffee and talked of birds and men, and suddenly a duck waddled in. She tilted her head sideways, looked at us, muttered

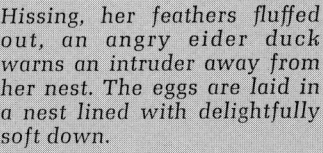

Hissing, her feathers fluffed out, an angry eider duck warns an intruder away from her nest. The eggs are laid in a nest lined with delightfully soft down.

softly to herself and crept underneath the sofa, where she had her nest. "That's why the door is open," Thoralf remarked matter-of-factly.

To the fishermen of Röst, as in other parts of northern Norway and on Iceland, eider ducks are welcome and lucrative guests – wild animals but so used to humans that one might call them semidomesticated.

Mollis in Latin means "soft," and the scientific name of the eider duck is *Somateria mollissima mollissima,* which one might loosely translate as "she of the super-soft body." After the gay days of spring courting are over, each eider duck builds a nest of grass and moss, then lines it with soft, greyish down plucked from her breast. As she lays her eggs, usually four to six, she adds more down and finally furnishes the nest with a "crown" of down, a sort of raised rim. She rarely leaves the nest during the incubation period, but when she does, she carefully pulls the soft crown like a coverlet over her eggs.

The fishermen harvest this eider down. They take the crown and part of the nest lining and this the ducks replace. Later, when the ducklings are hatched, the fishermen make a second collection, picking up the rather soiled down left in the nests. The down is dried and cleaned. In former days this was a laborious job. The down was rubbed by hand across great wire-mesh sieves. Now most down is cleaned with machines. Crowns and linings of twenty-five to thirty-five nests make a pound of cleaned down which sells, depending upon quality, for ten to twenty dollars. In Norway and Iceland, thousands of pounds of eider down are collected annually, but only slightly more than one thousand pounds in all of Canada.

(Left) While most of the females are home on the nest, the flashy eider drakes, their duty done, disport themselves in gay little bachelor groups. (Above) Eiders in flight.

Eider down is one of the poorest conductors of heat. A garment lined with down keeps body-warmth in, cold out. I have been cozily comfortable in an eider-down sleeping bag at 30 below zero; and the down is so elastic that a big, billowy bag can be rolled up and compressed into a small, compact bundle.

Eider ducks in Iceland and Norway are protected by law and, possibly even more effectively, by ancient custom. The Vikings built eider duck shelters. The discovery by Otto Sverdrup of Norse-type eider duck "houses" on St. Helena Island in the western portion of Jones Sound, between Devon Island and Ellesmere Island, has led some scientists to the conclusion that Vikings must have penetrated to this remote region of the Arctic. The Vikings also collected down from the immense eider duck colonies on islands off the West Greenland coast, and down was one of their important exports. (They also exported narwhal and walrus tusks, gyrfalcon and polar bear skins.)

The attitude of white settlers in North America and of Eskimos was built on the premise that the only good duck was a dead duck. The immense eider colonies along the south Labrador coast had already been decimated early in the nineteenth century by "eggers" from Halifax and Boston. Their method was simple and brutal. They travelled from island to island, smashing all eggs. A week or so later, they returned to collect the second clutches of eggs the birds had laid, thus ensuring themselves a reasonably fresh cargo. When Audubon visited the coast in 1833 he was horrified by the ruthless spoliation of the once-great colonies, through the pillage of eggs and massacre of birds, and predicted the extermination of the birds. He wasn't far wrong. The islands were made into bird sanctuaries in 1925, but during a Canadian Wildlife Service census in 1965, only eight thousand eider ducks were counted.

The Eskimos have little use for down. They collect the eider eggs and kill the birds. Until recently, about 100,000 eider duck eggs were taken annually in Greenland, and 150,000 or more ducks were shot.

On the Belcher Islands in Hudson Bay, the Eskimos make garments of the duck skins. It takes about twenty-five eider duck skins to make a *mitvin*, as a man's feather parka is called; and the woman's dress, the *amautik* is also made of duck skins, so that the baby in the voluminous hood sleeps in a cosy nest of down.

Recently, though, a new awareness of the ducks' potential value as down producers has led to their protection in some areas of the Canadian Arctic. At Port Burwell on Killinek Island, at the extreme northeast tip of Labrador, the Eskimos, prodded by an energetic administrator, began to collect down. When they found white men really did pay money for it, they suddenly took a new interest in the ducks of the area and passed and enforced among themselves some pretty stringent protective laws.

In 1900 only 2,000 greater snow geese came from the Arctic to the St. Lawrence River valley. Protection and management of the beautiful birds have increased their numbers to more than 50,000.

Protection and wise management have helped another arctic bird to increase in numbers. The greater snow goose nests on northern Baffin Island, and on Devon and Ellesmere Islands. In fall, nearly all the geese from these colonies fly to the Cap Tourmente area, east of Quebec City. In 1900 the flock was small – only 2,000 geese were counted. By 1921 they had increased to 6,000; to 20,000 by 1941; and in 1958, 47,500 greater snow geese were counted in this area. I once spent a day on the coastal marshes of Ile aux Oies (Island of Geese) in the St. Lawrence, just across from Cap Tourmente, and the great white geese covered the shore in such numbers that, from a distance, they looked like snow banks.

Their smaller cousins, the lesser snow geese, come in two editions, white and blue. Like the arctic fox, they are dimorphic. The blue goose was considered a separate species for a very long time, and its breeding grounds were one of the great ornithological mysteries of this century. The Canadian biologist J. Dewey Soper began his long search for the nest of the blue goose in 1923. He finally found it seven years later, after having travelled thousands of miles by boat and dogteam, on June 26, 1929, on a great plain on southwestern Baffin Island.

Geese of the blue phase seem to be on the increase lately, and there is some speculation this may be connected with the warming trend in the arctic climate. Being white has an obvious camouflage value in a snow-covered landscape, but if the snow disappears earlier than it used to, a white goose on the brownish-dark tundra is extremely conspicuous. Under these circumstances, it may be advantageous to be blue.

With powerful wingbeats a gaggle of Canada geese soars up into the air.

Three downy whistling swan cygnets on a tundra pond on Coats Island.

Unlike the gregarious snow geese who like to breed in great colonies, the Canada geese, whose range stretches from the Atlantic to the Pacific and right up to the Arctic Ocean, do not care much for company during their breeding season. They are wise and wary birds and, while they fill the autumn air with stirring honks during their migration, they are cautious and secretive near the nest.

Some 30,000 Canada geese congregate each fall at the famous Jack Miner Bird Sanctuary near Kingsville, Ontario. There the birds, so shy and elusive in the north, show little fear of man. They feed unconcerned a hundred yards from the road leading through the sanctuary, showing no fear of the thousands of people who come daily to watch them. About 3,000 Canada geese are caught and banded each year. In addition to serial number and address, each ring bears a Bible quotation. Each year the sanctuary administration receives from two hundred to four hundred letters from hunters who have shot banded geese. Some come from the Arctic, where Royal Canadian Mounted Police officers collect the bands. One missionary wrote apologetically that "his" Eskimos were unwilling to part with the bands. They were wearing them as amulets!

When I lived one spring with Pewatook on Jens Munk Island in northern Foxe Basin, his son Maliki shot some king eider drakes from the floe edge. The king eider is well named. It is a magnificent bird: the head a pearly grey, the cheeks a faint olive green, back and belly black, the neck and chest an immaculate white. Above its bill, the male carries a large, bright orange knob. It consists of a fatty substance and is regarded as a delicacy by the Eskimos. Maliki immediately bit off a knob and graciously offered me another, but I can't say I enjoyed it. It was like chewing on an oil-flavored eraser.

One of the ducks carried a band, requesting that it be sent to Copenhagen, Denmark. Back at our home, I showed Pewatook and Maliki that this duck had presumably come from Greenland. They were duly amazed, but refused adamantly to part with the band. I jotted down the information and sent it in, and was informed that this king eider drake had been banded three years previously in the Upernavik region of West Greenland.

Not very far from the Jack Miner Canada goose haven, where Long Point juts out into Lake Erie, bird lovers assemble each spring to watch the northward migration of one of the Arctic's most beautiful birds, the great white whistling swans. Years of protection have done them good, and they are again fairly numerous.

Once these great graceful birds were intensively hunted, in the south by white settlers, in the north by Eskimos and Indians. A New Englander, Thomas Morton, wrote in 1632: "There are of them [swans] . . . great store

The little whistling swan cygnets head across a meadow to the safety of their home lake. One of the wary parents (above) patrols the shores of the nesting island.

(Overleaf) Massed common murres stand shoulder to shoulder, finding security in numbers; marauding gulls avoid the center of such a large concentration of birds.

at the seasons of the yeare. The flesh is not much desired of the inhabitants, but the skinnes may be accompted a commodity, fitt for divers uses, both for fethers, and quiles."

The Hudson's Bay Company began to export swan skins from the north in the eighteenth century. Samuel Hearne noted at Churchill in the 1770s that the Indians were loath to bring in swans "the skins of which the Company have lately made an article of trade" because skinning the birds spoiled the flesh. Had it not been for this "thousands of their skins might be brought to market annually."

Soon they were. By 1810 the Hudson's Bay Company and North West Company were exporting each about 1,000 swan skins a year. In 1828 the H.B.C. sold 5,072 swan skins in London, and the next year 347,298 goose, swan and eagle quills were put up for sale. Between 1853 and 1877, more than 17,000 swan skins were sold in London by the Hudson's Bay Company.

Swan's down was made into powder puffs. Swan-skin lined muffs were the rage, and ladies' garments were trimmed with swan skin and ornamented with tippets of swan's down. The quills were greatly in demand. Audubon used quills of the trumpeter swan in drawing the feet and claws of small birds. These quills, he said, were "so hard, and yet so elastic, that the best steel pen of the present day might have blushed, if it could, to be compared with it."

In addition to commercial exploitation, the graceful swans were avidly hunted for "sport." By the beginning of this century the trumpeter swan had been nearly exterminated and the whistling swans had become rare. Decades of protection and care saved the trumpeters from extinction, and they number again about two thousand, while the whistlers in North America have increased to nearly one hundred thousand.

A pair of whistling swans had its nest at a lake on Coats Island, along the route of our twenty-mile polar bear trap line which we had to patrol each day. The nest, on an island in the lake, was inaccessible; but as we passed the lake, either cob or pen (as male and female swans are called) followed our progress attentively, sailing across the lake's dark water, the neck held ramrod-straight. One day, in July, I surprised both parents on a large meadow near the lake. They were moulting and unable to fly, but could they run! Whooping loudly (whistling swans don't whistle!), they raced flat-footed and fleet across the meadow, waving their wings, and out-distanced me with ease.

Some weeks later I met their downy cygnets on the same meadow, one big, one medium-sized and one small. The big one seemed to have the role of elder brother. It led its siblings into a little pond and they paddled busily away. When I approached the water's edge, they seemed to realize that the little pond offered hardly any protection and, after a short, worried, sotto voce palaver, they climbed out of the pool and headed single file, led by big brother, across the great green meadow to their large home lake.

(Top, left to right) The short stout bill and white gape mark identify the Brünnich's murre, which incubates its eggs on the ledges of great cliffs. Both adult and young murres are extremely noisy; a large colony is in perpetual uproar. With the egg pressed against a brooding patch on her belly, a Brünnich's murre sits high on a cliff on Coats Island.

(Below) Most common murres on the St. Mary Islands incubate their eggs in rock clefts. Off-duty parents congregate on favorite ridges. (Right) The "bridled" common murre, having a white eye-ring and stripe behind the eye, is a frequent genetic variation. In the foreground is the typically pear-shaped murre egg. Because of its shape it rolls in a very small circle and is less likely to fall off the ledge.

A few miles from swan lake, on the great granite cliffs of northeastern Coats Island, was the home of two great colonies of Brünnich's murres (or "guillemots" as they are known in Britian). These murres of the Arctic, and their cousins of the more temperate regions, the common murres, are, with the possible exception of the starling-sized dovekies, or little auks, the most numerous sea birds of the northern hemisphere. The Canadian ornithologist Dr. Leslie M. Tuck has estimated their total population at about sixty million.

One day in early October, as I was travelling in northern Baffin Bay aboard the Hudson's Bay Company ship *Pierre Radisson*, the mate suddenly called me. "What do you think of this?" he asked, and showed me several extensive marks, flashing greenish-white on the radar screen. "Ice fields?" I guessed. "Keep looking!" he suggested. After a while I realized my ice fields were changing their position. I looked at the mate, puzzled. "Birds," he said, with a grin.

Later we saw them, black clouds flying low above the sea, bunching and spreading – tens of thousands, maybe hundreds of thousands of murres on their southward migration.

(Above) Brünnich's murres on the nesting cliff. (Right and below) Common murres. (Facing page) Colonies of common murres covering acres on a northern island; in the background an iceberg floats past in the Labrador Current. Common and Brünnich's murres winter offshore. (Overleaf) Brünnich's murres crowd every available ledge on their nesting cliffs on Coats Island.

Dovekies in the Thule district of northern Greenland. Dovekies breed in Greenland, Iceland, Spitzbergen and some other arctic areas but are not known to breed in Canada. They breed on talus slopes in colonies which may number in the millions. They tend to gather on the snow patches on the breeding slopes in dense groups.

The largest murre colony in Canada is on Digges Island, and on the nearby mainland cliffs in the vicinity of Cape Wolstenholme, where Hudson Strait and Hudson Bay meet. These colonies have been known for a long time. Henry Hudson stopped at Digges Island in 1610, to restock his food supply with murres, before continuing the exploration of the great bay to the south. He spent the winter locked into the ice of James Bay while his famished crew dreamt of big fat juicy murres. In spring he meant to continue his search for a passage to Cathay, but his men had murres on their mind, abandoned Hudson and a few faithful crew members and set sail for Digges Island and birds. This time they were not the only ones to hanker for murre meat. Eskimos were at the bird cliffs, there was a dispute and several of the mutineers were killed.

When I visited Ivugivik, the settlement closest to Digges Island and Wolstenholme, more than three-and-a-half centuries after this event, the Eskimos still preserved a vivid oral tradition of their first and fatal encounter with white men, and could tell me in great detail what had happened. "Why did the Eskimos attack the whites?" I asked Tomassi Mangiok. "They killed them from fear," he said. "Those white men looked terrible to them." Then he added, ingenuously: "We are more used to them now."

The murres, in common with other arctic and sub-arctic alcids, have brown-black backs and white undersides. It's an elegant costume and an eminently practical one. The white feathers have air-filled cells, and form a protective cushion against the icy water on which the birds spend the

Dovekies. These birds winter at sea, ranging south into the North Atlantic and the Baltic Sea.

greater part of their life. The dark feathers favor the absorption of the sun's warm rays.

Murres don't bother to build a nest. They just plonk their single, pear-shaped egg onto the bare rock of a cliff ledge, roll it onto their big, flat, webbed feet and brood it against a bare, warm patch on their belly. The parents alternate in this task, and very carefully pass the eggs from one partner to the other.

The cliffs are high, the ledges narrow and crowded – the birds sit so close they look, from a distance, like beads on a string – there is a continual coming and going, jostling and fighting, and in all this brouhaha about half of all the eggs laid roll off the ledges and splatter on the rocks below.

The percentage of lost eggs would probably be much higher if the eggs were not ingeniously designed to minimize such mishaps. They are pear-shaped and, consequently, when they roll they describe only a fairly small arc. As incubation advances, the egg's pointed end becomes progressively heavier. The blunt end, which becomes lighter, tilts up and the circle described by a rolling egg becomes smaller, a vital adaptation to the frequently extremely narrow, and sometimes outward sloping, breeding ledges.

A murre colony is technically referred to as a bazaar, and there is something apt about it. Each murre is at least as noisy as a haggling housewife, and that, multiplied by a few hundred-thousand, can be deafening. Once the chicks emerge they add their shrill, hungry cheeping to the general racket, and the adult birds buzz to and fro on narrow busy wings, to haul food for the greedy, fast-growing youngsters. It has been estimated the murres of the world gobble up at least 25,000 tons of food each week, or 1,300,000 tons of fish and planktonic animals per year.

It is hard to imagine how they can manage, unless one is aware of the stunning fecundity of the arctic seas. One day, I rowed across the great Saglek Fiord in northern Labrador. Suddenly I saw below me in the water, about three or four feet from the surface, a solid layer of fish, millions upon millions of capelin, all facing in the same direction. I slowly rowed across the fiord, and wherever I looked, there was this greyish-speckled stratum of fish, square miles of them. And this was but one shoal in one fiord. Yet capelin and the small polar cod occur in similar swarms throughout the seas of the northern hemisphere, a nearly inexhaustible bounty for hungry murres.

When the time comes for the young murres to go to sea, those born on flat islands, led by adults, march to the water on foot. Those born on cliffs must jump, although they are still unable to fly, in some cases from a height of seven hundred or even a thousand feet. They are, understandably, not too keen about it. The adults assemble at the foot of the cliff and call encouragingly. On the ledges, the youngsters waddle back and forth, peer into the void, and recoil in fright. Finally the most courageous, or hungriest, jumps. It falls, it planes, its wings flap wildly, and usually one or two adults accompany it in its spectacular descent. In the water the young murre is surrounded by a mob of milling adults and, after a while swims out to sea with one or more of them, not necessarily its parents. It seems that with this plunge from the cliff the parent-child bond is severed, and henceforth the little murres are part of the larger murre community.

Common murres prefer a somewhat milder climate than do the hardy Brünnich's murres. In Europe they have colonies as far south as Spain. They usually nest on cliffs, or on the surface of suitable, but flat, islands. One peculiar colony is on the St. Mary Islands, near the north shore of the Gulf of St. Lawrence. Here the stratified and inclined rock is frequently rent by broad fissures, some six or seven feet deep, and down at the bottom the murres incubate their eggs. There they are safe from marauding gulls; but in some years melt water seeps into the clefts and inundates them. Murres can shuffle along very slowly with the precious egg held between feet and belly, and they will advance thus on the inclined slope of the cleft above the water. On this steep and slippery slope the egg exchange between parents becomes a real juggling feat, and the guardian of the sanctuary told me that in such water-filled clefts most eggs are lost. I walked past the fissures, and from their depth came a steady stream of squawk and squabble. Occasionally some murres scrambled out. It was obviously filthy down there; one looked like a gentleman in tails onto whose immaculate shirt front a waiter has just accidentally emptied a cup of cocoa.

They jostled and crowded each other on a favorite ridge. A few flew off to feed and returned, breasts and bellies shiny white again, braked above the ridge, with wings and webbed feet, tried to find a place to land, and plopped down haphazardly amongst their complaining fellows. After a while duty called and, scrabbling and flapping their wings, some murres slithered down a cleft, the plutonian chorus below picked up volume, greetings, insults and eggs were exchanged and a new relay of mucky murres emerged from the deep, badly in need of food and a bath.

In Newfoundland, along the Labrador coast and in the English-speaking villages along the north shore of the Gulf of St. Lawrence, different, local, names are used for most sea birds. Little auks are "bull birds," black guillemots, "sea pigeons." And when I asked a fisherman at Blanc Sablon, near the Strait of Belle Isle, what birds nested on little Perroquet Island, he said: "Puffins, tinkers and turrs" – puffins, razor-billed auks and murres.

The typical mass-flight behavior of the dovekies. When a predator, such as a gyrfalcon, approaches, the dovekies take flight in a tightly packed group. The predator is confused and is incapable of making a choice, with the result that often no kill is made at all.

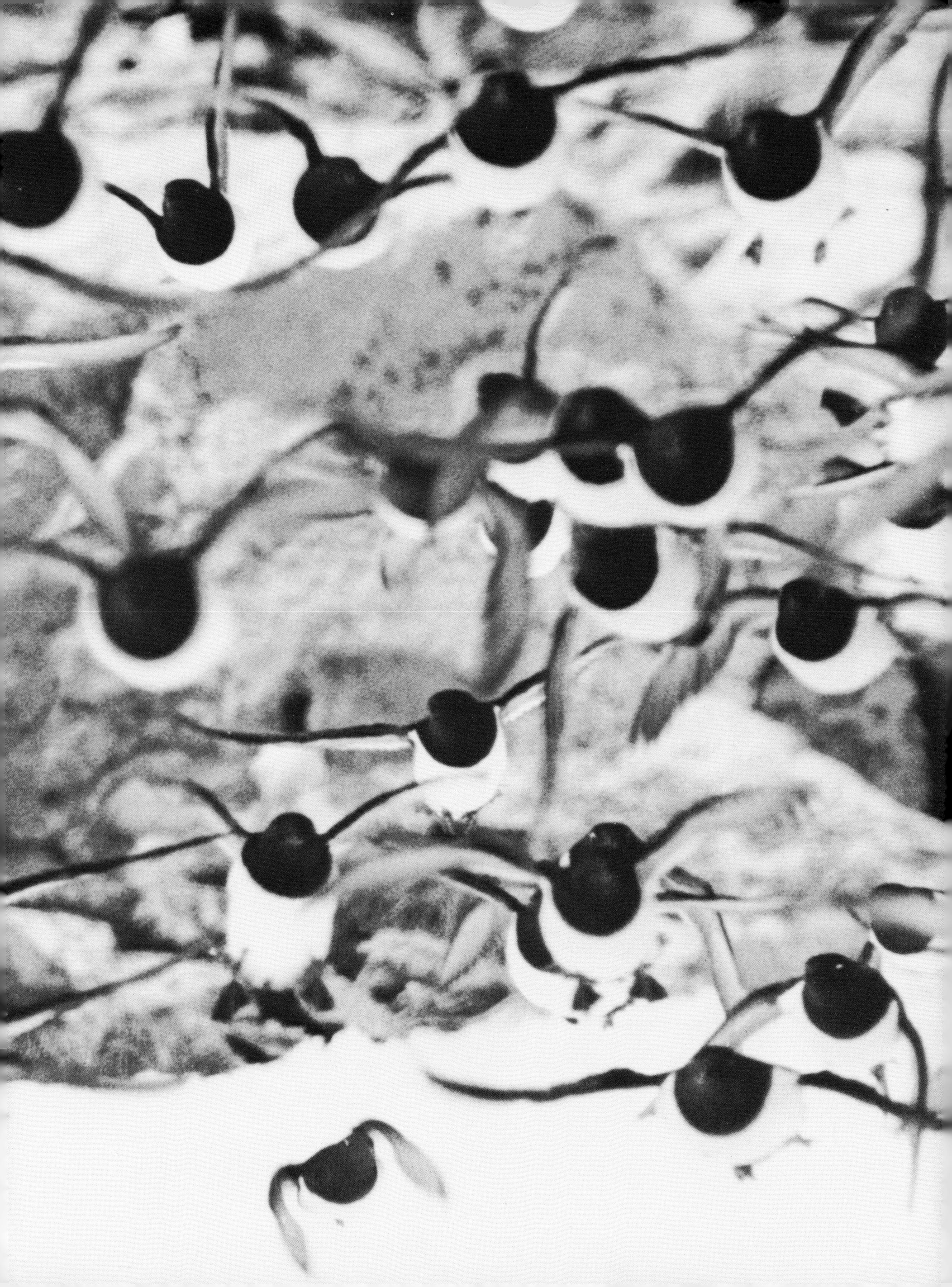

The Sea Unicorn

Once, while not busy painting, building fortifications or designing such futuristic vehicles as the tank or the helicopter, Leonardo da Vinci put his mind to the problem of how to capture the wary unicorn. This animal, by all accounts rare, shy and elusive, had one known weakness. It liked girls. "In its lack of moderation and restraint and the predilection it has for young girls, it completely forgets its shyness and wildness," wrote Leonardo da Vinci. "It puts aside all distrust, goes up to the sitting girl and falls asleep in her lap. In this way hunters catch it."

Had they really existed, trapping unicorns would have been a lucrative occupation since their great horns were worth in medieval times at least their weight in gold. The horns were considered a panacea, good against any ailment from ague to plague, and particularly prized for their reputed ability to detect and neutralize poison. Since the possibility of being poisoned was a professional hazard for princes and potentates of that age they willingly paid fortunes to obtain unicorn horns.

Considering Catherine de'Medici's later reputation for ruthlessness, it was thoughtful of her uncle, Pope Clement VII, to present her father-in-law with a unicorn horn when she married the dauphin of France, towards the middle of the 16th century. Charles V, Holy Roman emperor, settled a hefty debt with the Margrave of Bayreuth by giving him two unicorns, and an electoral prince of Saxony paid one hundred thousand thalers for his unicorn horn. Charles I of England also had such a horn. It may have saved him from poison but was no help against the axe.

When Robert Dudley, Earl of Leicester, favorite of Queen Elizabeth I, demanded in 1576 from the Warden and Fellows at New College, Oxford, their treasured unicorn horn, they were understandably reluctant to part with it. Finally a compromise was reached. They gave Leicester the tip and kept the rest. The truncated horn is still at Oxford. Like most of the other "unicorn horns" that have survived from medieval times, it is the canine tooth of the narwhal, the small arctic whale having in the male a straight, spiraled, and tapered ivory tusk up to ten feet in length.

But in those days few people had heard of a narwhal and everyone believed firmly in the unicorn. Authorities like Aristotle and Pliny had

Jutting from the left side of the narwhal's head, the spiraled tusk is really an enormously elongated canine tooth. The tip is often broken, showing the hollow pulp cavity.

vouched for its existence, and the thirteenth century English monk Bartholomaeus Anglicus, relying in part on Pliny, described it thus in his book *De Proprietatibus Rerum:* "An unicorn is a right cruel beast... shaped like to the horse in body, and to the hart in head, and in the feet to the elephant and in the tail to the boar... and [has] an horn strutting in the middle of the forehead of two cubits [three foot] long."

This composite beast may sound bizarre but it wasn't entirely imaginary. The Chinese have always prized rhinoceros horn, partly for the medicinal virtues they ascribed to it, but mainly as a supposedly very effective aphrodisiac. Their demand for rhinoceros horn lasts to this day, and has been largely responsible for the near-extermination of rhinoceroses in India and East Asia. Distorted tales of this wonder-working horn drifted to Europe, got tangled up with descriptions of the oryx, the Arabian antelope whose great curved horns looked in profile like one horn, and with the undeniable existence of those superb spiraled narwhal tusks offered for sale by shrewd traders as "unicorn horns."

These horns were one of the most valuable items of the medieval pharmacopoeia. They were made into goblets in which poisoned wine was supposed to froth, and into eating utensils used by the kings of France until 1789. Since a pinch of unicorn horn was worth a lot of money, adulterations were common. As one medieval writer complained: "Some wicked persons do make a mingle-mangle thereof... being compounded of lime and soap or other which things to make bubbles arise, and afterwards sell it for the unicorn's horn."

Narwhal tusks, whose real origin, one suspects, was a closely guarded commercial secret, reached Europe by two main routes. Once the Vikings became established in southwestern Greenland at the beginning of the eleventh century, they made long trips northward to hunt sea mammals. The runic inscription left by such a hunting party was discovered near Upernavik, and scientists assume Vikings travelled at least as far north as Melville Bay. Thus they roamed nearly the entire west coast of Greenland, probably the world's best narwhal area, and they must have hunted them avidly, because narwhal tusks became one of Greenland's most valuable exports. Icelandic records state that a ship with a cargo of "unicorn horns" was wrecked off the Iceland coast in 1242, and Denmark's famous "unicorn throne" was constructed of narwhal tusks.

Other tusks came from Siberia. Collected by Yakuts, Koryaks, Chukchees or Eskimos, they passed from tribe to tribe, either south to China or west to Russia. In China, some of the tusks, called *tu-na-si,* were fashioned into sword or dagger hilts. Others were traded to India and a few of these eventually reached Europe, usually via Arabia. Other tusks probably came to western Europe from Russia. When the ghost-haunted Tsar Boris Godunov sent a delegation to Shah Abbas of Persia to conclude

Narwhal on the beach at Koluktoo Bay.

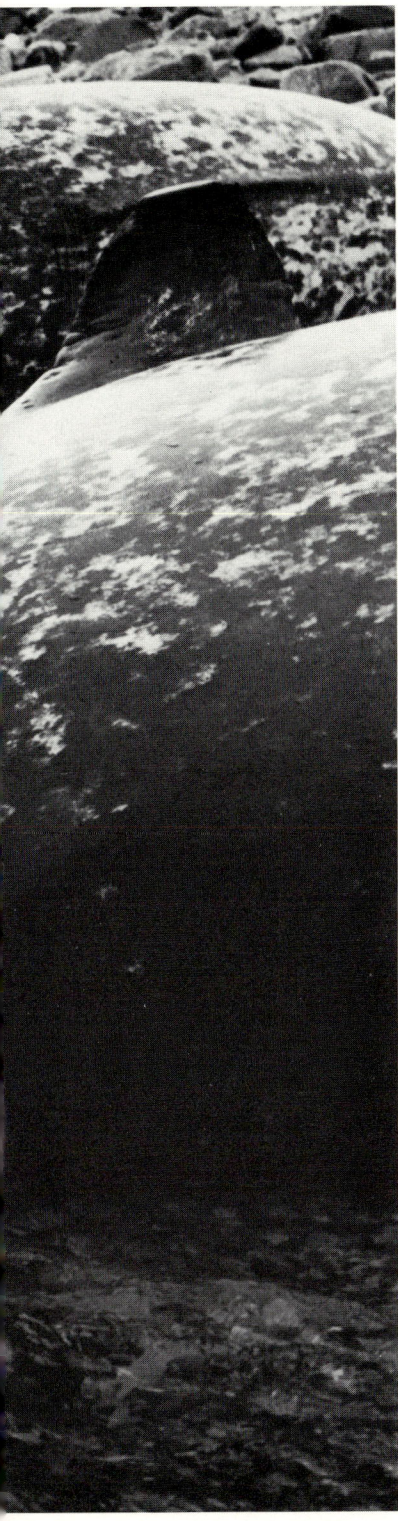

an alliance against the Turks, he thoughtfully made his fellow monarch a gift of seven "fish teeth."

But while Europeans were quite familiar with the narwhal's tusk they knew nothing about the animal that carried it and clung firmly to their belief in unicorns. Even the "Father of Zoology," Conrad Gesner, author of the famous work *Historia Animalium* published in 1551, though frankly sceptical about unicorns, was forced to conclude: "But one has to trust the words of wanderers and far-going travellers, for the animal must be on earth, or else its horn would not exist." And he printed a picture of a stag-like unicorn from whose forehead sprouts what is unmistakably a narwhal tusk.

It was not until the age of arctic exploration that Europe got its first description of the narwhal. On his second voyage to Baffin Island, in 1577, Martin Frobisher and his crew found in Frobisher Bay "a dead fishe floating, whiche had in its nose a horne streight & torquet of lengthe two yardes lacking two ynches, being broken in the top, where we might perceiue it hollowe, into whiche some of our Saylers putting Spiders, they presently dyed ... by the vertue whereof we supposed it to be the sea Unicorne." The "spider test" was commonly employed to check the genuineness of unicorn horns. Since spiders were supposedly poisonous and unicorn horn "killed" poison, its touch was deemed deadly to spiders. The narwhal tusk was taken back to England and "... is to be seene and referued as a Jewel, by the Queens maiesties commandement, in hir Wardrop of Robes."

Arctic whalers soon became familiar with the narwhal, but they spun such fabulous yarns about the fierceness of the sea unicorn for the benefit of awed home audiences that even the famous French naturalist Comte de Buffon believed the tales. The narwhal, he wrote, "revels in carnage, attacks without provocation ... and kills without need."

The narwhal (*Monodon monoceros,* meaning "one-tooth one-horn"), a remote cousin of the dolphins (its only close relative is the beluga or white whale), is a small, gregarious whale of the circumpolar arctic seas. It is most common in Baffin Bay and Davis Strait, and occurs locally in northern Hudson Bay, particularly in the Repulse Bay area. It is seen north of Norway, European Russia and Siberia, in the region of Spitsbergen, Franz Josef Land, Novaya Zemlya and Severnaya Zemlya, all the way to Wrangel Island, but seems to be extremely rare in the Chukchee and Beaufort Seas. While Nansens's famous ship, the *Fram,* drifted with the ice through the central polar basin, its crew saw narwhal at nearly 85 degrees north, and the odd narwhal strayed as far south as New England, Britain and Holland. One narwhal ran aground in 1684 on the Island of May, in Scotland's Firth of Forth, and was described by Tulpius as the *unicornis marinum*. Although schools of narwhal numbering over two thousand individuals have been

A narwhal is caught in the broad-meshed net, its tusk gleaming through the clear water.

seen, their total number in the Canada-Greenland section of the Arctic is estimated at about ten thousand, and a somewhat smaller number may inhabit the rest of the arctic seas.

The narwhal probably owes its name, derived from the Old Norse *nahvalr,* "corpse whale," to its peculiar coloration. Young animals are a dark, slaty blue-grey. Adults are mottled with a mass of dark dots and dashes upon a greyish-white background, merging into nearly solid black areas on the back, while the underside is nearly pure white. The older the narwhal are, the lighter their color.

Adult males reach sixteen to twenty feet in length, the females are somewhat smaller. Males weigh between one-and-a-half and two tons. The body is sleek and spindle-shaped, and terminates in elegantly twin-lobed flukes. The males, in particular, are rather bluff-browed and their small pectoral fins are coquettishly up-curled. The eye is small and sloe-shaped, and the otic opening behind it smaller than a pinhead, but it leads to a complex and extremely sound-sensitive inner ear.

What makes the narwhal so remarkable is the great, straight, twisted and tapered tusk jutting from the head of the male. It is really an enormously elongated canine tooth. Both male and female narwhal have canine teeth (and no others), but in the female, with very rare exceptions, these remain rudimentary and stay hidden in their alveoli. In the male, only the left canine develops into a tusk, invariably sinistrally spiraled. Occasionally twin-tusked males have been killed. In these the right tusk is usually shorter than the left one, and it, too, has a left-handed twist.

The tusks can measure up to ten feet in length with a basal girth of about eight inches. They consist of high-grade, fine-grained ivory, are hollow (the pulp-filled dental cavity extends nearly to the tip) and quite brittle and fragile. The tip is usually a brightly polished creamy white. The rest of the tusk, particularly the grooves of the spirals, is covered with a brownish-green algal growth. The tips of tusks are frequently broken and, since the tissue near a fresh break is often inflamed and sometimes putrid, one can assume that such an unfortunate narwhal suffers a ten-foot tooth ache.

Ever since the narwhal became known, sailors and scientists have pondered over the use the narwhal may make of his formidable tusk. In *Moby Dick,* Melville mentions some of the theories of his day. ". . . some sailors tell me that the Narwhale employs it for a rake, in turning over the bottom of the sea for food. Charley Coffin said it was used for an ice-piercer; for the Narwhale rising to the surface of the Polar Sea, and finding it sheeted with ice, thrusts his horn up, and so breaks through. . . . My own opinion is, that however this one-sided horn may really be used by the Narwhale . . . it would certainly be very convenient to him for a folder in reading pamphlets."

Since then other theories have been propounded: that the males use the tusks like rapiers to fence for the favor of their tuskless females, or that they use the tusks to skewer fish. Since their favorite food, (in addition to skate, squid, polar cod and cuttlefish) seems to be Greenland halibut, the Arctic explorer Peter Freuchen has suggested narwhal use their tusks to prod up these bottom-dwelling fish, a theory seconded by Eskimo lore. Now, most scientists think that the tusk is merely an ornament, a secondary sexual characteristic of the male, like the mane of a lion or a cock's comb. Presumably due to their lopsided dental development, narwhal are among the rare animals to have a completely asymmetrical skull.

Narwhal spend the winter either in the open sea south of the ice, or in "polynias," large areas of open water in the high Arctic, kept ice-free by strong currents and up-welling water. They usually swim together in small pods of from ten to twenty animals. Peary, on one of his polar trips, encountered them and wrote: "A school of narwhal dashing to windward, their long white horns flashing out of the water in regular cadence, and the waves dashing jets of spray from their bluff foreheads"[1] – a lovely description of fast-swimming narwhal who then often raise their tusks above water when they surface. When swimming leisurely, they usually come up in a smooth rolling motion; one sees the head and part of the dark back with its inch-high dorsal ridge (narwhal and belugas lack dorsal fins), hears a resoundng "swoooosh" as the animals exhale from their crescent-shaped spiracles, and down they dive again, usually in unison·

In early summer, narwhal congregate in large schools and seek out the pack ice near the coasts. Here they are safe from their worst enemy, apart from man, the killer whale. Freuchen says killer whales hunt narwhal in packs and kill them by ramming them with great force. As the ice begins to break up, narwhal swim into favorite fiords and bays of Greenland or the Canadian arctic archipelago. The Canadian ornithologist Dr. Leslie M. Tuck, studying murres on Bylot Island, off northern Baffin Island, saw a mass migration of narwhal. They swam past the island at the rate of about three hundred an hour, roughly two thousand three hundred of them in the course of one long arctic day, heading for Eclipse Sound and Milne Inlet.

The calves are born in any month of the year, though June and July seem to be the favorite months. Narwhal are a slow-reproducing species; it is believed females bear only one calf – very rarely two — every three years.

Summer and fall are spent in bays, fiords and inlets. Sometimes narwhal linger there too long. Ice forms across the entrance of bay or fiord, and the narwhal, who have to surface at least every fifteen or twenty minutes to breathe, are trapped. Slowly, inexorably, the ice closes in upon them, while the animals jostle each other in the shrinking space of open water to reach the life-giving air. Eskimos call such schools of ice-trapped whales (narwhal

A sleek, tuskless female narwhal.

or belugas) *savssats*. Freuchen wrote: "We have seen cases where more than a thousand narwhal have been trapped this way and have died" and added, "One can hear the sound the unfortunate animals make for miles."[2] The Danish scientist Morton P. Porsild was in Greenland in the winter of 1915, when a sudden, severe frost laid a barrier of ice across Disko Bay and trapped large numbers of narwhal. He estimated the Eskimos killed about one thousand of the whales.

As long as the hole is large, the narwhal surface in their usual rolling motion, breathe, and dive again. But their own vapor-laden breath, settling on the rim of their gelid prison as ice, hastens its contraction and soon the trapped animals can only breathe in the constricted space by surfacing nearly vertically, their tusks jutting high above the ice – looking from a distance, said one observer, like chevaux-de-frise.

To find a savssat is, obviously, the dream of every Eskimo. The dark meat of narwhal is excellent. Eskimos eat the blubber, or render it into oil to burn with a smokeless yellowish flame in their crescent-shaped soapstone lamps. Narwhal sinews once were used as thread, and the skin was cut into stout thong, ideal for dog team traces since it remained supple, even when it had been wet and dried out again. (This quality also made narwhal skin a desirable trading item. Ski bindings were made of it.) But, above all, the Eskimos prize the nearly inch-thick skin, called *muktuk*, as a delicacy. The raw, rubbery skin has a bland, slightly nutty flavor, and is extremely rich in vitamin C, containing nearly as much of it per unit of weight as oranges or lemons.

But unless narwhal are trapped in a savssat, they are hard to hunt. They are wary, and extremely fast swimmers. As long as Eskimos approached them nearly noiselessly with their sleek kayaks, they might harpoon the odd one. Now Eskimos use motor-driven canoes, and narwhal flee long before they are near them. Since narwhal navigate by sonar-like echolocation, they deftly avoid the nets Eskimos set to catch seal, but they are taken in special large-mesh whale nets, which do not seem to reflect a sufficient echo to warn the animals of danger. The Eskimos' best chance is still to hunt narwhal in early summer, when they surface in the leads among the pack ice. There, as Freuchen describes it, the narwhal "...run backwards and forwards in small companies from six to ten and bob up in all directions; sometimes they lie still for a space of nine or ten minutes and 'pugsignerpoq,' by which is understood the remarkable halt often made by whales, at the same time bobbing up and down, and then suddenly dive all together as if at a signal." The Eskimos kill about one hundred to one hundred and fifty narwhal in the Canadian Arctic each year and a similar number is taken along the Greenland coast, though it may be considerably higher in years when Eskimos discover large schools of narwhal trapped in savssats.

In addition to harboring a host of endoparasites, narwhal have one rather peculiar ectoparasite: the narwhal louse. It really isn't a louse at all, but a highly specialized little crustacean (*Cyamus monodontis*) who clings to his sleek and fast-swimming mount with sharp-hooked little legs. One usually finds a few at the base of narwhal tusks, and dense clusters in old wounds or skin abrasions.

Although the narwhal's "miraculous" tusk has been known to Western man for nearly a thousand years and the animal itself for about four hundred years, much about its life, movements and migrations is still shrouded in mystery and conjecture. The only intensive study of narwhal was undertaken by Dr. Arthur Mansfield of the Arctic Biological Station of the Fisheries Research Board of Canada.

Eskimos in Pond Inlet, on northern Baffin Island, suggested Koluktoo Bay, near the head of Milne Inlet, as a good area to catch and study narwhal. In early August, they told Dr. Mansfield, narwhal come into this bay in vast numbers, swimming close past the great promontory of Bruce Head.

Twin-tusked narwhal are rare. This specimen with two tusks of equal length is especially unusual.

The Eskimos were certainly right. In the summer of 1964, Dr. Mansfield and two technicians, Brian Beck and David Robb, were standing high on Bruce Head when, suddenly, the narwhal came. From their vantage point, they could see the animals swim close together, more than two thousand of them, the tusks of the great males flashing through the greenishly pellucid water.

A year later I flew into Koluktoo Bay ("bay of waterfalls," in Eskimo) with Brian Beck and David Robb, to catch the arctic unicorn. We set four great nets, each a hundred yards long, twenty yards deep, with eighteen-inch mesh, along the coast of the bay and then spent frantic days saving them, when large fields of ice drifted majestically and menacingly towards them.

In previous years the narwhal had arrived with great regularity in the early days of August. Now the ice seemed to keep them away. Each day we patrolled the nets, crossing over such vast schools of char we could feel the turbulence they created in the water through the pliable bottom of our canoe. But the narwhal did not come.

We were not the first to wait for whale in Koluktoo Bay. Near our camp, in a small and amazingly lush valley dotted with yellow buttercups and mauve willow herb, were the remains of several Thule culture Eskimo houses. The walls were built partly of stone, partly with huge bones of Greenland whales, and in the house depressions lay still the whale ribs that had once served as roof rafters for the dwellings of these early whale hunters. On the shores of Koluktoo Bay we often found the great, porous bones of the bowhead whales (as whalers called the Greenland whale). In their lee, gaining both shelter and nutrients from the disintegrating bones, bloomed bright-yellow arctic poppies.

Suddenly, on August 13, the narwhal were in our bay, swimming singly, or, more often, in pods of up to twenty animals. They arrived about the same time at the head of Milne Inlet. At the mining camp there, the cook, who had never heard of narwhal, dashed excitedly into the mess hut and shouted: "There's a big fish in the bay and it plays with a long stick!"

For us hectic days followed: disentangling captured whales from the clinging mesh, cutting them up, collecting organ samples, caching all the meat, skin and blubber for the Pond Inlet Eskimos and crawling at night exhausted into our sleeping bags, reeking of formaldehyde and blubber.

Strangely enough, although the whales seemed sleek and well fed, with a thick blubber layer, none had food in their stomachs. Once four great males, each one measuring over fifteen feet in length, with tusks six foot long, got tangled up in the same net. We hauled them ashore with block and tackle at high tide and returned early next morning when the tide was out. Now the narwhal lay high and dry, looking so strange with their massive, twisted tusks one could well understand that for ages people had

At Koluktoo Bay, Thule Eskimos hunted the great Greenland whales long ago. In the lee of their massive bones, gaining shelter and nutrients, grow arctic poppies.

preferred to believe in a stag-like unicorn, rather than in the existence of an ivory-tusked whale in the remote seas of the Arctic. Nor can one blame the mechanic who, newly arrived in a northern settlement, saw some Eskimos haul a great male narwhal ashore. He walked over, looked at the tusked whale in incredulous wonder, and finally said: "There ain't no such animal!"

We captured seventeen whales, a welcome bounty for the Pond Inlet Eskimos and their dogs. On one of the last evenings we were sitting in the tent, when we heard the resounding "swoooosh" of a surfacing narwhal. We looked out and saw the animal heading straight for the net closest to camp.

David and I ran over. It was a superb evening, calm for once, and the sea a sombre bronze in the afterglow of the setting sun. As we stood on a cliff above the net, the narwhal hit it, and began its silent, desperate struggle for life, deep down in the dark water. The great styrofoam floats bobbed up and down in a macabre jig of death, spreading concentric ripples across the burnished surface of the placid sea. After fifteen minutes, a stream of glistening bubbles pearled towards the surface, rested there an instant, popped and vanished. The sea unicorn was dead, and we returned to camp, silent and thoughtful.

The narwhal's closest cousin is also an arctic whale, the white whale, or beluga, a word derived from the Russian *belyi*, meaning white. Most belugas live in the far north, but one quite isolated group make their home in the St. Lawrence River where Jacques Cartier spotted them on his second voyage to Canada in 1535. In the vicinity of Ile aux Coudres, he "discovered a species of fish, which none of us had ever seen or heard of. This fish is as large as a porpoise but has no fin. [Like the narwhal, the white whale lacks a dorsal fin.] It is very similar to a greyhound about the body and head, and is as white as snow, without a spot upon it. Of these there are a very large number in the river, living between the salt and the fresh water. The people of the country call them *Adhothuys* and told us they are very good to eat. They also informed us that these fish are found nowhere else in all this river and country except at this spot."

The Indians at the time of Jacques Cartier were catching white whales off Ile aux Coudres. More than four hundred years later a copy of their ingenious whale trap was still employed near the island to catch white whales for the aquaria of the world.

I was a bit further upriver, in the ancient village of Tadoussac, when a fisherman told me: "You should go to Ile aux Coudres. They have caught two *marsouins blancs* in their trap and they are going to bring them to the aquarium in Quebec City." *Marsouin* is the French-Canadian name for white whale. It is rather unflattering. It comes from the Norwegian and means "sea pig."

When I arrived at the quay the ferry from the island was just making fast, and off rolled a big truck carrying an improvised tarpaulin basin. In it, lolling in three feet of water and protesting loudly, were two belugas, a snow-white mother and her bluish-grey calf. White whales are notoriously noisy. They squall, growl, grunt, chirp, click, scream and squeal, and they can whistle so loud, sailors used to call them "sea canaries." The poor whales on the truck where in no whistling mood, but each time the vehicle lurched the calf squeaked shrilly and was answered by a fairly sotto voce grunt from its mother.

After seeing the whales off to the aquarium, I went to the island to have a look at the trap that had caught them, the last whale trap in the world. Once there were many along the St. Lawrence. The biggest and most deadly efficient was at Rivière-Ouelle. In the late nineteenth century, it caught up to five hundred belugas in a single day. At Ile aux Coudres, formerly surrounded by seven of these strange traps, 320 whales were taken in one tide. Trapping white whale was the most important industry on this isolated island. In his *History of New France*, published in 1720, the Jesuit Father Charlevoix says: "The white porpoise yields a hogshead of oil.... they [the islanders] make puddings and sausages of their guts; the puck is excellent fricasseed and the head preferable to that of a sheep. The skins are tanned and dressed like Morocco leather.... they shave it down until it becomes a transparent skin.... fit for making into waistcoats and breeches, it is always excessive strong and musket-proof. Nothing exceeds it for covering coaches."

Now the giant trap was only used to catch belugas alive, to be shipped to aquaria in North America and Europe. One whale caught in this trap was flown to England. The islanders took me out to see their singular whale weir. In spring they cut about four thousand young trees, two inches thick and fifteen foot high. With some of these they build a sort of fence from the island two miles out into the river. They ram the poles, three to ten feet apart, through a thin layer of sand down into the soft, clinging clay of the river bottom. The kidney-shaped trap is three miles long and two miles wide. Like the fence, it is built of poles, set in the shape of a gigantic C. From the lower lip of the C the fence leads to the shore. The upper lip is curved inward toward the trap.

The belugas pursue shoals of fish upriver with the rising tide. As the tide turns, so do the whales, leisurely following fish and current downriver. This brings them to the fence and, although it consists mainly of large gaps, the whales dare not traverse it. The strong current makes the pliant poles shiver and shake, and to the whales' sonar-like echo-location system the row of spaced palings must appear like a solid wall. They swim alongside

Eskimos at Whale Cove capture white whales in nets, then haul them aboard their boats.

the fence and this brings them to the great maw of the trap, the open portion of the C.

Once in the trap, the whales swim parallel to the poles, following the outline of the enclosure. They swim around and around in the great weir, not daring to pass between the quivering poles, until the outgoing tide leaves them wallowing helplessly in shallow water where they are easily caught.

In 1861, P. T. Barnum, the circus king, heard tales of whales caught in the St. Lawrence. His showman instincts aroused, he rushed north, hired thirty-five French-Canadian fishermen, and they did catch two white whales. The belugas were packed in great crates with seaweed and rushed to New York, where Barnum announced to an awed populace how he personally had made every arrangement "for capturing and keeping alive two of these monsters." A tank was built for the "leviathans" and "they are now swimming in the miniature ocean in my Museum, to the delight of visitors." The delight and the whales were short-lived. They died a few days after their arrival.

The belugas were once so common in the St. Lawrence, they were considered a threat to the fisheries and a bounty was paid until 1939 for their destruction. Since then a new realization of the whales' beauty has resulted, if not in their protection, at least in a greater tolerance. Fishermen now proudly point out vantage points along the St. Lawrence to tourists, where they can admire the sleek white whales, as they surface in sociable and garrulous clusters.

The white whales in the St. Lawrence are considered a relic population. In the last post-glacial period, when the great ice sheets that had covered much of Canada gradually retreated northward, the St. Lawrence was an immense fiord. Belugas caught fish where Montreal now stands, and roamed as far as Ottawa. Gradually the fiord shrunk to the river of present, still mighty proportions, and its white whales became isolated from the stocks of belugas that retreated with the ice to the seas of the far north.

They are common along the west coast of Greenland, among the islands of the Canadian arctic archipelago, along the western arctic coast of Canada and in Hudson Bay. The inhabitants along the coasts of Russia's White Sea hunted white whale as early as the ninth century and paid part of their tribute to the Princes of Novgorod in beluga hides. Now the Russians catch large numbers of white whales in great steel nets, but they are still numerous north of the Soviet Union. Even in recent years schools of over ten thousand white whales have been seen to migrate past Novaya Zemlya.

After whalers in the nineteenth century had nearly exterminated the great Greenland whale they augmented profits by killing thousands of white whales in the arctic seas. Near Cape Sparbo, on Devon Island, one whaler killed 700 belugas in a single day. Whole fleets of ships from Norway pursued them relentlessly in the Spitsbergen area and thousands more were taken in Hudson Bay.

In Cumberland Sound, on Baffin Island, the Hudson's Bay Company operated a white whale fishery for many years. When the whales came into the sound in early summer, Eskimos in canoes and whaleboats, making a tremendous racket and shooting onto the water just in front of the whales, stampeded whole pods into shallow water where the helpless animals could be easily killed. Their hides were made into driving belts and boot laces, and the great sooty trypots in which whale blubber was rendered into oil still stand in the village of Pangnirtung. The Eskimos on Greenland also knew how to herd schools of white whale into shallow water and Freuchen said in Godthåb Fiord "several hundred of them could be taken in this way in one day."

The white whales spend the winter in the open sea, or in ice-free areas of Hudson Bay and the high Arctic. Early in spring they travel northward, seeking out the broken pack or swimming along the coasts, where food is plentiful and they are relatively safe from the powerful and ferocious killer whales. If killer whales come upon a pod of beluga in the open sea, they dive below the hapless white whales and ram them with such force that they can throw a three thousand-pound male several feet out of the water, bursting its stomach and breaking its ribs. To avoid this fearful fate, belugas may seek the safety of the pack ice, where killer whales with their tall, scimitar-shaped and sensitive dorsal fins are loath to follow. Vilhjalmur Stefansson, crossing the Beaufort Sea from Alaska to Banks Island in 1914, was amazed to see white whales as early as May 21. ".... [we] saw a school of beluga whales northward bound along the lead. During the next two or three weeks we saw thousands of them. They were usually travelling north or east according to the way the leads were running...." Sometimes heavy frost iced over all open water. "Then it was interesting to see the six- or eight-inch ice bulge and break as they struck it with the humps of their backs. A moment after the noise of breaking ice would come the hiss of the spouting whale and a column of spray."[3]

In summer large schools of white whales seek out fish-rich estuaries of favorite rivers, like the Mackenzie River in northwestern Canada or the Churchill River, flowing into Hudson Bay. Although they have been hunted at Churchill for centuries, and a large settlement has grown up on the banks of the river, the white whales arrive each year, in early summer, undeterred by the whine of outboard motors or the massive drone of the great freighters that come to load wheat from the towering rows of grain elevators. I sat on that quay once on a rather blustery day in early July and watched the whales, rolling up in the turgid water, sending little puffs of vapor into the chilly air as they exhaled with loud snorts, and disappearing again into the silt-laden stream.

A white whale is hauled aboard at Whale Cove. The skin, or muktuk, will be canned in piquant sauce for the Southern gourmet market.

A few hundred yards away stood the whaling plant, where some of these whales would be turned into oil, and mink-food and belts. Big-game hunters would come later in summer and pay hundreds of dollars for the "thrill" of killing the whales with high-powered rifles, keeping their jawbones as trophies with which to impress friends back home. Another school of whales surfaced near the dock, grunting in a self-satisfied sort of way, and dived again into the murky water, and I wondered what powerful atavistic urge drives men from the comforts of their cities to the shores of Hudson Bay for the sole purpose of killing these beautiful white whales.

At least the days of indiscriminate and unlimited slaughter are past. Quotas limit the kill, and scientists from the Fisheries Research Board of Canada are studying the mystery of white whale migrations to provide a basis of knowledge onto which rules for management and conservation can be grafted. In recent years teams of Indians, working for the Fisheries Research Board, have driven large numbers of white whales into the shallows of Seal River, north of Churchill, so scientists could mark them with tags. A few of the marked whales were captured within a few weeks several hundred miles further north, in the Repulse Bay area.

Like narwhal, belugas sometimes linger too long in the bays and fiords of the north and become trapped by fast-forming ice in fall. In the winter of 1966, an Eskimo from Grise Fiord on Ellesmere Island, returning from a hunting trip, found a white whale savssat in Jones Sound, forty miles east of the settlement. During December and January, when the moon was full and temperatures dropped to forty and fifty below, the men from Grise Fiord came to kill the luckless whales. It was easy. The whales could not escape. To breathe they had to surface in the dark pool of water, already less than forty feet across. The Eskimos harpooned and shot the whales and hauled them home as dog food. The belugas had been trapped since October and none of the whales killed by the Eskimos had food in their stomachs. They seemed to live off their fat reserves, and already the blubber layer of many animals had shrunk to less than half its normal thickness. About 80 of the 150 trapped whales were taken. The rest the Eskimos left alone, to struggle for life and air in the bluish light of the arctic night. No one expected the whales to survive. Throughout the long winter, Eskimos visited the savssat from time to time, and each time the hole was smaller.

The surviving, emaciated whales were jostling each other for a vital space to breathe, the vapor of their breath settling as ice on the rim of their prison. When I arrived in Grise Fiord in April, the whales had been trapped for nearly six months. Some time earlier an Eskimo had seen them. The hole was now less than a yard across and nearly completely roofed by a cupola of ice. When I drove out with one of the Eskimos, the whales had gone. A tidal crack had broken the ice, and along this fissure the surviving whales had presumably escaped to open water. During the many months

A female narwhal surfaces at Koluktoo Bay and dives with a splash.

of their imprisonment in the savssat, these whales had apparently survived by absorbing their own blubber which, in a well-fed adult beluga, can attain a thickness of about eight inches.

The Eskimos have, of course, always hunted the white whale. Its meat is eaten, raw or cooked, used as dog food or is dried in thin strips to serve as provision for winter. The rendered blubber produced an oil burning with a nearly white flame in the Eskimos' crescent-shaped lamps. In former times, meat and blubber were cached to let them ripen and, according to Freuchen, when the blubber was "green as grass in spring" it tasted just marvellous. Oil rendered from the liver is rich in vitamin A, and the thick skin, called *muktuk,* has the same sort of hazelnut flavor as narwhal skin and, like it, contains considerable amounts of vitamin C.

In Whale Cove, on the west coast of Hudson Bay, I accompanied Eskimos who netted white whales. The great, wide-meshed nylon nets were set at strategic spots among a maze of rugged islands and inlets, a favorite area of the whales, and many got tangled up in the greenish mesh and drowned. Most of the meat was used as dog food. Some of it was sent to the settlement of Rankin Inlet, to be canned and sold in other Eskimo villages. In the cannery Eskimo women, using their traditional *ulus,* crescent-shaped knives, cut the white whale muktuk into neat cubes. Packed in a piquant sauce and canned, this Eskimo delicacy was destined for the gourmet market in the south.

Throughout the northern seas, about five thousand white whales are still killed each year, for their oil, their skin and their meat. Yet despite this persecution, the beluga seems to hold its own. It is certainly rarer in the Spitsbergen area than it was in former years, and the schools in the St. Lawrence seem smaller than in the past, but in much of its arctic realm, migrations of hundreds and sometimes thousands of these whales can still be seen. Once, flying in a small plane along the west coast of Hudson Bay, I saw a pod of more than a hundred belugas swim purposefully northward. Their heart-shaped flukes rose and fell in graceful, undulating rhythm, propelling the sleek torpedo-like whales, glistening in immaculate whiteness in the dark waters of Hudson Bay.

The Lazy Shark

We thought the shark was dead. It hung limp and apparently lifeless in the great net we had set in Koluktoo Bay to catch narwhal. Brian Beck and David Robb, with whom I worked on the Fisheries Research Board narwhal study program, hauled up the grey, inert form. We freed the shark from the net, careful not to touch it with our bare hands, since its rasp-like skin is covered with sharp denticles. If one gets cut by a shark's skin, a rather nasty rash or inflammation can result. It was easy to get the shark out of the net. It was held only by a few meshes and, once caught, had apparently devoted little effort to freeing itself. We put a rope around the base of its tail and towed the shark into the shallows near our camp, where I could photograph it at leisure.

We wrenched open its low-slung mouth, and admired the shark's impressive and efficient dental equipment: rows of small, sharp teeth in the upper jaw, and a single, serrated blade of a lustrous pearly white in the lower jaw. Greenland Eskimos used to make saws out of it. Later we saw what the sharks can do with these teeth. They went after the dead narwhal hanging in our nets, and scooped out circular gobs of skin and blubber, shearing off some ten pounds with each bite. The cuts through the whale's taut skin were as smooth as if they had been made with a razor.

To get pictures of the teeth, Brian suggested we keep the jaws ajar with a stick. While David and I opened the mouth, he shoved in a short, stout stick and, just when he had it nicely in place, the jaws clamped shut with a chilling chomp. Brian flicked his hand out in time. The stick stayed in. "Muscular reflex," said Brian, his imperturbable calm only slightly shaken. We got the jaws open again and Brian gingerly adjusted the stick. This time the jaws snapped shut even faster, and a shudder ran through the great grey shape.

"You know," said Brian, "I think he's still alive." As if to emphasize those words, the shark rolled over, made a few slow and languid movements with its tail, headed for David, who side-stepped with alacrity, and gently and without haste our erstwhile captive left us for the deep water of the bay.

The well-armed, low-slung mouth of the Greenland shark, small teeth in the upper jaw, a saw-edged plate in the lower jaw.

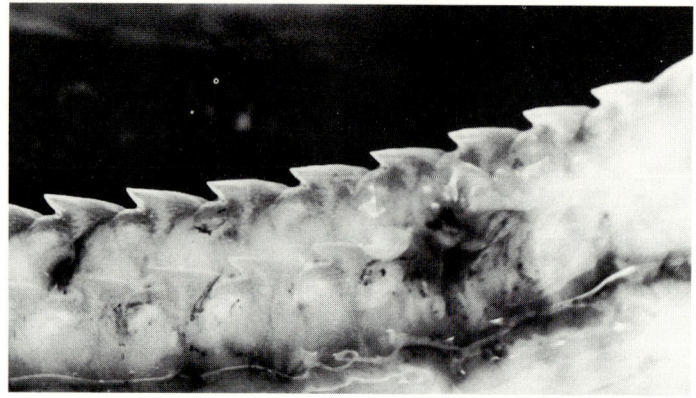

It was a Greenland shark. There are more than three hundred species of shark. Most of them prefer the warm or temperate oceans of the world, while the Greenland shark has rather successfully occupied an ecological niche in the boreal seas. Apart from this, it is mainly known for its lethargy. It is often called "sleeper shark," and its scientific name rather stresses the point that this shark is neither speedy nor brainy – *Somniosus microcephalus* – "the somnolent one with the tiny head." "No doubt this shark is the laziest fish known in the Arctic," was Freuchen's judgment.

Lazy and lethargic as it seems, the Greenland shark may nevertheless come up with the odd burst of speed, for there has been found in its stomach the remains of various fast-swimming fishes, such as char, as well as of seals; and it would take a fairly fast shark to catch a seal. Normally, though, the shark seems to be predominantly a carrion feeder. Like other sharks, he is attracted by blood and, where whales were cut up in the Arctic, Greenland sharks have suddenly congregated and gorged themselves so voraciously that they would let themselves be killed rather than be driven away from the food.

Adult Greenland sharks can weigh over a ton and attain a length of more than twenty feet. There is some indication that the Greenland shark may be a veritable marine Methuselah. A shark caught and tagged in 1936 off the Greenland coast was recaptured in 1952. In these 16 years, the shark had only grown slightly more than three inches, from nine foot one inch, to nine foot four-and-a-half inches. If this growth-rate is in any way typical, a fifteen- or twenty-foot shark may be centuries old. No wonder they are not in a hurry. They've got all the time in the world!

Whales and seals store energy reserves as blubber that surrounds their bodies like a thick blanket. On this they can draw in times of dearth. Sharks store their fat reserves in the liver. The liver of a well-fed Greenland shark can weigh over six hundred pounds and consists of up to 60 per cent oil. This oil is extremely rich in vitamin A, containing about two thousand to four thousand international units per gram.

This, and the fact that the Greenland shark is clothed in a scratchy but tough and valuable hide, has led to its wholesale slaughter. Canadian Eskimos are rather leery of the sharks and, although they are quite numerous in many areas of the Canadian Arctic, have never caught them, either for themselves or on a commercial basis.

The Greenland Eskimos, too, feared the shark. Freuchen says this was due "to the Eskimos' belief that the sharks somehow were able to think, almost like human beings and that they were capable of avenging each other's death."[1] The promise of monetary gain obviously overcame the Eskimos' scruples. Once there was a market for shark products they began to catch them: a few thousand a year early in the nineteenth century, eleven thousand to fifteen thousand a year by the end of that century and about

(Facing page) A close-up of the peculiar, worm-like parasite found attached to the corneas of nearly all Greenland sharks, and a close-up of the efficient serrated cutting edge of the lower teeth, fused into a single band. (Right) Sharks have cut great chunks of blubber from this narwhal carcass, shearing it out in large circular gobs.

fifty thousand annually in the 1940s and 1950s.

The sharks are caught on baited longlines, with steel cable or chain leaders, or in winter through holes in the ice. They seem to be a deep-sea fish, staying normally at one hundred fathoms or deeper, but their acute sense of smell will lead them to the alluring bait, and if it has already been taken by a fellow shark they go ahead and eat him. Freuchen says fifty thousand sharks will yield about ten thousand barrels of liver from which sixty thousand pounds of oil could be extracted. After being relieved of its vitamin content it is sold as lubricating oil. Long ago, it was quite a cheap oil, and known as *trekronetran* – "three kroner oil," in Denmark.

The skin is made into a tough, durable leather, called shagreen. Sold in squares, complete with its highly abrasive denticles, it is used by cabinetmakers as a sort of long-lasting emery paper. Most of the hides, though, are run through heavy steel rollers that press the denticles into the skin, giving it an attractive speckled pattern. This leather is sometimes called "galuchat," after Jean-Paul Galuchat, master glove-maker to Louis XV of France, who used shark leather to cover the scabbards of the king's ceremonial swords. Now the leather is used for handbags and wallets, and in bookbinding.

Fresh shark meat is toxic. If dogs eat it, they become dizzy and delirious, and also get diarrhea. If they eat too much of it, they may die. Ravens or fulmars, who may gobble up pieces of fresh meat when the Eskimos cut up sharks are similarly affected. They have trouble flying, croak and squawk and flop around, until the poisonous effect wears off. But once the meat is either dried or partly putrified it is quite all right to eat, provided one cares for high shark meat. In Iceland, where Viking descendants hunted Greenland sharks for centuries, it was customary to bury the shark meat for a while. It was then thoroughly washed before being cooked. The

A great Greenland shark in shallow water at Koluktoo Bay.

Greenlanders usually cut the meat in thin slices, dry it, and use it as dog food, mixed with seal blubber and meat. Canadian Eskimos seem to know only that it is toxic, but have apparently never learnt that it loses its toxicity, due to an extremely high ammonia content, with time.

To judge from the numbers alone of sharks caught along the Greenland coast, they must be regarded, despite all their apparent indolence, as a successful and well-adapted species. Apart from man they are not known to have any enemies; the arctic seas are rich in food, even for a sluggish carrion grubber; they may live to an immensely ripe old age; and they seem to procreate at a pretty good rate. A pregnant female may carry a barrelful of shell-less, hen-sized eggs, but whether she is oviparous or viviparous has not yet been determined. Since a close cousin, the warmth-loving *Somniosus rostratus* of the southern Atlantic and the Mediterranean is viviparous, it is assumed the Greenland shark may also hatch its eggs inside, and then give birth, rather than spawning them.

Although its natural realm is the arctic seas, Greenland sharks occasionally come in considerable numbers southward – as far as Japan and Oregon in the Pacific and down to Scotland and France; and they have been taken repeatedly off the New England coast. Since the Greenland shark is predominantly a scavenger, a sort of floating garbage collector, and is particularly numerous near fish plants and whaling stations, it is known in some areas as the "gurry shark."

The Greenland shark has the dubious distinction of being host to one of the most peculiar parasites in existence. It looks like a whitish worm and makes its home on the shark's eyeball. Called *Ommatokoita elongata*, it is really a crustacean, an immensely specialized copepod. In a study made in Greenland, 1,505 sharks were examined: 84 per cent carried a parasite on each eye, 14.5 per cent were infested on only one eye and only 1.1 per cent were parasite free. The few without copepods may have been only temporarily vacated by these parasites, since their eyeballs bore tiny circular scars, presumably made by previous occupants.

The copepods vary in length from about 3 mm to 70 mm and attach themselves firmly with a small circular disk to the shark's cornea. Normally sharks carry one on each eye, but occasionally several cling to one eye. One would think it rather vexing to have a tenant on each eyeball, but the sharks don't seem to mind and there is even a theory they may profit from this only apparently unpleasant association. It has been suggested that the yellow-white, worm-like copepods serve as lures. The shark itself is dark grey and hard to see, while the copepods dangling from its eyes flash alluringly through the water. When a fish comes along, all eager to snap up what looks like a tasty morsel, it finds itself eaten by the shark instead. It seems like a nice symbiotic arrangement to make life easy for an indolent shark.

Lake of the Cannibal Cod

In the late 1930s, reports reached fisheries experts that Eskimos were catching cod in a small lake near the south coast of Frobisher Bay on Baffin Island. They were puzzled. Cod were not supposed to exist in Frobisher Bay.

It is true, cod have been moving northward in recent decades. In the last century, during warm spells in 1820 and between 1840 and 1850, they had reached Greenland's west coast in appreciable numbers. Then cold weather set in again, and the cod stayed away. In 1913 only five ton of cod were caught along the entire Greenland coast. Since then, as the arctic climate has warmed up, they have increased rapidly, so that cod fishing is now one of Greenland's major industries.

At Port Burwell on Killinek Island, at the extreme northeast tip of Labrador, a cod fishery has been started within the last decade and when I visited the settlement in 1968, Eskimos were hauling up cod by the boatload. Cod have also been caught a bit further north, near the bleak Resolution Island, off the southeast tip of Baffin Island. But not one cod has ever yet been seen in Frobisher Bay.

The mystery remained unsolved until 1951, when Dr. Max Dunbar of Montreal's McGill University, working in Frobisher Bay from the Fisheries Research Board vessel *Calanus*, visited Ogac Lake (*ogak* is Eskimo for "cod"). The lake lies at the head of a five-mile fiord, Ney Harbour, and is connected with it by a roughly one-hundred-yard-long cascading river.

Tides in Frobisher Bay are among the highest in the world, and at peak tides this river is reversed and flows from the sea into the lake. The lake's surface water is sweet to a depth of about fifteen feet. From there to its bottom, the lake is filled with salt water, and little or no interchange takes place between the water layers. The salt-water layers have about three-quarters the salinity of Atlantic Ocean water. Close to the lake's bottom the water is stagnant and samples taken by Dr. Dunbar smelled strongly of hydrogen sulphide.

Ogac Lake is L-shaped, the longer arm reaching about a mile inland, the shorter one measuring roughly three-quarters of a mile. It is surrounded by beautifully rugged mountain ranges, soaring two thousand feet and more above the lake, snow-streaked even in summer. Two small rivers, one of them called Ogac River, the other still nameless, flow into the lake, and their sweet water seems to glide above the heavier salt-water layers to the river that normally rushes down into Ney Harbour. A sill in the lake, at the outlet of the river, allows fresh water to run off, but retains the salt water of the deeper layers.

The cod live in the salt-water layers of the lake. Scientists think these fish got stuck in the lake during a warmer climatic period, maybe one thousand years ago (the time the Vikings established settlements on Greenland) or possibly as long as four thousand years ago. Since the land in much of the eastern Arctic has risen considerably, in some places as

much as six hundred feet in the past ten thousand years, it is possible that Ogac Lake was not always a lake, but formed once a prolongation of the Ney Harbour fiord.

At any rate, while other cod retreated southward, the Ogac Lake cod were stuck, landlocked and famished in a food-poor lake. They eat seaweed, and to this they may owe their peculiar golden-brown color. They gobble up the spiky sea urchins. And when neap tides reverse the river from sea to lake, they cluster near the sill and greedily grab the sculpins washed in from the fiord. But that only happens a few times each year, and in between these rare feasts there is prolonged famine. So, it seems, the cod are driven to their last resource: they eat each other. They are so voracious and cannibalistic that Dr. Dunbar has said: "If a seal got into Ogac Lake, I believe it would be torn apart by the cod."

And so it might, because cod in Ogac Lake grow to an impressive size. A subsequent McGill University expedition caught one cod 4 foot 5 inches long and weighing 56 pounds, and another 4 foot $7^{1}/_{2}$ inches long. They have the peculiar shape of fish on a poor diet. The heads, always big in cod, are even more massive in the Ogac Lake cod, while the bodies, in comparison, appear meager and spindly.

It is easy to catch the landlocked cod. I visited Ogac Lake with some Eskimos from the settlement of Frobisher, a day's canoe trip away. We spent the night camped near the river, got up at high tide and roped our canoe through rocks and rapids up into the lake. We had found an old jigger on shore and had a couple of fishing rods along. The moment the lures trundled down through the crystal-clear waters of the lake, the cod bit. They were ravenous. Iola, beside me, pulled in a small cod. Suddenly a great brownish shape rose out of the deep, white mouth agape, and grabbed the smaller cod. Happy, the Eskimo tried to haul in his double catch. The big fish rose reluctantly, eyes bulging, straining against the line. He was nearly in the boat, when his mouth opened, the little cod jerked out and the great fish disappeared.

The same thing happened to the scientists who visited the lake. When they pulled up a small cod, a big one surged up to grab and swallow it. Many fish were tagged, and released. Some of the tags of smaller cod were later recovered – from the stomachs of bigger cod. Locked into this food-poor lake, the cod fall back on each other, to lead a cod-eat-cod existence.

Ogac Lake is not unique. Landlocked cod have been caught in Mogilnoe Lake on Kildin Island in the Barents Sea; in lakes on Bering Island in Bering Sea and on Bolshoy Shantar Island in the Okhotsk Sea, as well as in a lake near the coast of the Kamchatka Peninsula. With the exception of Mogilnoe Lake, which contains both sweet and salt water, these are all fresh-water lakes. In none of them do the landlocked cod attain nearly the size of those great, voracious, golden-hued cod of Ogac Lake.

(Facing page) The Ogac Lake cod have large heads and rangy bodies. (Above) Arctic char ascend the glittering waters of a swift stream on northern Baffin Island.

If the Ogac Lake cod is a fascinating accident of climate and geography, the fish really important to the north is the sleek elegant arctic char. In some regions they occur in tremendous numbers. In Koluktoo Bay, northern Baffin Island, when we went to patrol narwhal nets, we sometimes passed over schools of char so dense that the water boiled and the canoe bounced as they scattered at our approach.

Unlike salmon, who first see life in fast-flowing streams, char are spawned and spend the first three or four years of their life in arctic lakes, as far north as Ellesmere Island and northern Greenland. Some char never leave their home lakes. They are the domestic type and remain a bit stunted and runty in comparison to their more enterprising cousins who, once they have attained maturity, are gripped by wanderlust. They swim down to the sea early in spring when the ice breaks up–in May in the low Arctic, in June further north. For the next two months they gorge themselves on capelin or polar cod, and put on weight nearly visibly. They usually stay fairly close to the river they have descended, hunting mainly near its mouth or in the adjacent bay or fiord. In July, fit and fat, they are ready for the return journey to their home lakes.

The males are lovely now, head and back black-brown, with a dark greenish sheen, the sides flecked with bright orange spots, the belly a bright red. To this they owe their name, which comes from the Gaelic *ceara* meaning "blood-red." They ascend the rivers and, although they can swim up some strong rapids, they cannot hurtle themselves up over great waterfalls like their cousins, the salmons. Most of the char have reached their lakes by August or early September, and spawn there.

It is during the migration from the sea to the lakes that Eskimos in some areas used to catch them in large numbers. They built stone weirs, called *sapotit*, and harpooned the massed fish in the eddies below. In winter, too, they speared char through holes in the ice, luring them into range with little, fish-shaped ivory decoys.

Now most char are caught with nets. The most important char fishing grounds are in the fiords of Labrador, particularly north of Nain, where Eskimo fishermen hauled in nearly two hundred thousand pounds of char in 1968, enough to fill more than eight hundred big barrels. Char are only fished in the fiords, permitting enough to run the gantlet of nets, reach their lakes and provide for a new generation of char. This way up to a thousand barrels, each one holding about 200 to 220 pounds of char, can be taken each year, and catches have been remarkably consistent for nearly a century: 579 barrels in 1883; 787 barrels in 1893; and 798 barrels in 1903.

Char is one of the few Eskimo delicacies that most white men truly appreciate, and commercial fishing has been started in many parts of the Arctic. The fish are frozen and shipped south, some are even canned, and the Eskimos are discouraged from feeding the valuable fish to their dogs.

I was sitting high on the great promontory of Bruce Head, in Koluktoo Bay, one day, hoping to witness the narwhal migration – to see hundreds, perhaps thousands, of the fabled sea unicorns rush through the glass-clear water. But the narwhal did not come. A couple of red-throated loons circled on the calm water. A ringed seal popped up, saw the loons, dived and mischievously came up underneath one of them, nearly scaring the feathers off him. Annoyed, the loons left. The seal disappeared, too, filling up on shrimp, probably. And then came the char, in dense shoals, like great, dark clouds underwater, bunching and scattering, swimming steadily westward. Within a day or two, they would reach their goal, the rushing, rippling Robertson River, and swim up its swift-flowing waters to the lakes far inland. They were safe now. But long ago, Eskimos had waited for them, spears poised, on the rock shelves jutting out above the eddies along the river banks.

We had found the remnants of an ancient Eskimo fishing camp near the rapids. They had taken char from the river, whales and seals from sea, and caribou and hares, ptarmigan and geese from the land. The Arctic had been rich then and there was a natural balance between its wealth and the people who lived off it. That balance was broken when the white man moved in to exploit, often ruthlessly, the animal wealth of this harsh land. But the basis for the Arctic's great natural wealth is still there: a sea, incredibly rich in nutrients; a land, infinitely vast yet unsuited for agriculture, which can provide bountifully for those animals that have chosen the Arctic as their ecological home. With management and protection, with knowledge and foresight, the great animal wealth of former days could once again flourish in its great arctic realm.

Accuse not Nature,she hath done her part;
Do thou but thine.
               John Milton, *Paradise Lost*

Notes

Death on the Ice
1. Wilfred T. Grenfell, *Down to the Sea, Yarns from the Labrador* (Fleming H. Revell Company, New York, 1910)
2. In a paper submitted in 1971 to the annual meeting of the International Commission for the Northwest Atlantic Fisheries, Dr. David E. Sergeant stated: "The combined sustainable yield (of Front and Gulf herds) at present is estimated to be no more than 125,000 young harp seals. Present annual catches average 218,000 young, and effort is increasing."

    In 1972 new legislation and international agreements finally limited the kill close to the sustainable yield figure of Dr. Sergeant. Only "landsmen" in boats less than 60 feet long may kill harp seal pups in the Gulf (in the past, the average landsman's kill has been 30,000 harp seals per year). A quota of 120,000 limits kill on the Front (60,000 for Canada, 60,000 for Norway).

"Bestes a la grande dent"
1. Fridtjof Nansen, *Farthest North* (Harper and Brothers, New York and London, 1903)

Nanook — The Great White Bear
1. C. R. Harington, "Polar Bears and Their Present Status," *Canadian Audubon* (January/February, 1964), published by Canadian Audubon Society, Toronto.
2. Vilhjalmur Stefansson, *The Friendly Arctic* (Macmillan, New York, 1921). Permission granted by MacIntosh and Otis, Inc., N.Y.

The Little White Fox
1. Charles Elton, *Voles, Mice and Lemmings: problems in population dynamics* (Clarendon Press, Oxford, 1942)

"Ill shapen beast"
1. Samuel Hearne, *A Journey to the Northern Ocean* (Pioneer Series), edited by Richard Glover (Macmillan of Canada, Toronto, 1958).
2. Otto Sverdrup, *New Land,* translated by Ethel Harriet Hearn, 2 vols. (Longmans, Green and Company, London, 1904).
3. J. W. Tyrrell, *Across the Sub-Arctics of Canada* (William Briggs, Toronto, 1908).
4. Stefansson *(op. cit.)*

The Living Barrens
1. Fridtjof Nansen and Robert Collett, "An Account of the Birds" in *The Norwegian North Polar Expedition 1893-1896* (Longmans, Green and Company, London, 1900)
2. Tyrell *(op. cit.)*
3. Hearne *(op. cit.)*
4. Ernest Thompson Seton, *The Arctic Prairies* (International Universities Press, New York, 1911).
5. Quoted by Fraser Symington in his *Tuktu,* published by the Canadian Wildlife Service.
6. Alfred E. Brehm, *Wildlife and Scenes in Many Lands* (Blackie and Son, London, 1895).

"A ger-fauk that is milke white"
1. Hearne *(op. cit.)*

The Sea Unicorn
1. Robert E. Perry, *The North Pole* (Hodder and Stoughton, London, 1910).
2. Peter Freuchen and Finn Salomonsen, *The Arctic Year* (Copyright © G. P. Putnam's Sons, New York, 1968). Reprinted by permission at the publishers.
3. Stefansson *(op. cit.)*

The Lazy Shark
1. Freuchen and Salomonsen *(op. cit.)*

C.N.F. BOOKSTORE F.C.N.
46 RUE ELGIN ST.
OTTAWA CANADA
K1P 5K6

Aug 6. 19 74

SOLD BY	C.O.D.	CHARGE	ON ACCT	ACCT. FWD.	
1 Encounters with Arctic					
2 Animals				15	25
3					
4					
5					
6					
7					
8					
9					
10					
11		TOTAL		15	25
12					

94478

94478

Bibliography

Banfield, A. W. F. *The Barren-Ground Caribou.* Canadian Wildlife Service. Issued under the authority of the Minister of Resources and Development, 1951.

Bent, Arthur C. *Life Histories of North American Diving Birds.* Dover Publications, New York, 1963.

Bent, Arthur C. *Life Histories of North American Gulls and Terns.* Dodd, Mead, New York, 1947.

Bent, Arthur C. *Life Histories of North American Wild Fowl.* Dover Publications, New York, 1951.

Biggar, Henry P. *Samuel de Champlain,* 6 vols. Champlain Society, Toronto, 1922-1936.

Biggar, Henry P. *The Voyages of Jacques Cartier.* Ottawa, 1924.

Birket-Smith, Kaj. *The Eskimos.* E. P. Dutton, New York, 1935.

Bö, Olav. *Falcon Catching in Norway.* Universitetsforlaget, Oslo, 1962.

Bruemmer, Fred. *The Long Hunt.* The Ryerson Press, Toronto, 1969.

Buckley, John L. *The Pacific Walrus.* U.S. Fish and Wildlife Service, Special scientific report no. 41, 1958.

Budker, Paul. *Whales and Whaling.* Harrap, London, 1958.

Cade, Tom J. *Ecology of the Peregrine and Gyrfalcon Populations in Alaska.* University of California Press, Berkeley, 1960.

Clyde, Kennedy. "Cannibal Cod in an Arctic Lake." *Natural History.* The Magazine of the American Museum of Natural History, February 1953.

England, George A. *Vikings of the Ice.* Doubleday, Page & Co., New York, 1924.

Elton, Charles. *Voles, Mice and Lemmings; Problems in Population Dynamics.* Clarendon Press, Oxford, 1942.

Freuchen, Peter. *Arctic Adventure.* Farrar & Rinehart, New York, 1935.

Freuchen, Peter. *Book of the Eskimos.* World Publishing, Cleveland and New York, 1961.

Freuchen, Peter. *Vagrant Viking.* J. Messner, New York, 1953.

Freuchen, Peter, and Salomonsen, Finn. *The Arctic Year.* Putnam, New York, 1958.

Godfrey, W. *The Birds of Canada.* National Museum of Canada, Bulletin no. 203, Ottawa, 1966.

Hakluyt, Richard. *Hakluyt's Voyages.* Edited by Irwin R. Blacker. Viking Press, New York, 1965.

Harington, C. R. *Denning Habits of the Polar Bear (Ursus maritimus Phipps).* Canadian Wildlife Service Report. Series no. 5, 1968.

Harington, C. R. "Polar Bears and Their Present Status." *Canadian Audubon Magazine,* 1964.

Harrington, Richard. *The Face of the Arctic.* Abelard-Schuman, New York, 1952.

Harrison, Richard J., and King, Judith E. *Marine Mammals.* Huchison University Library, London, 1965.

Hearne, Samuel. *A Journey from Prince of Wales's Fort in Hudson's Bay to the Northern Ocean 1769, 1770, 1771, 1772.* Edited by Richard Glover. Macmillan, 1958.

Hickling, Grace. *Grey Seals and the Farne Islands.* Routledge & Kegan Paul, London, 1962.

Jonkel, C. J. *Life History, Ecology and Biology of the Polar Bear, Autumn 1966 Studies.* Canadian Wildlife Service Progress Notes no. 1, 1967.

Jonkel, C. J. "Life History, Ecology and Biology of the Polar Bear in Canada." In *Polar Bear Research and Management in Canada.* Report to international meeting of polar bear specialists, Morges, Switzerland, 1968.

Jonkel, C. J. *Polar Bear Research in Canada.* Canadian Wildlife Service Progress Notes no. 13, 1969.

Kelsall, John P. "The Migratory Barren-Ground Caribou of Canada." Canadian Wildlife Service, 1968. Cover title: *The Caribou.*

King, Judith E. *Seals of the World.* British Museum (Natural History), London, 1964.

Ley, Willy. *The Poles.* Time Inc., 1962.

Lockley, R. M. *The Seals and the Curragh.* Dent, London, 1954.

Lockley, R. M. *Grey Seal, Common Seal.* André Deutsch, London, 1966.

Lubbock, Basil. *The Arctic Whalers.* Brown, Son and Ferguson. Glasgow, 1955.

McLaren, Ian A. *The Economics of Seals in the Eastern Canadian Arctic.* Fisheries Research Board of Canada, Arctic Unit, Montreal, 1958.

McLaren, Ian A. "Methods of Determining the Numbers and Availability of Ringed Seals in the Eastern Canadian Arctic." *Arctic,* vol. 14, no. 3, 1961. Arctic Institute of North America.

Macpherson, Andrew H. "The Abundance of Lemmings at Aberdeen Lake, District of Keewatin, 1959-63." *The Canadian Field-Naturalist,* vol. 80, no. 2, 1966.

Macpherson, Andrew H. "The Barren-Ground Grizzly Bear and Its Survival in Northern Canada." *Canadian Audubon Magazine,* January/February 1965.

Macpherson, Andrew H. *The Dynamics of Canadian Arctic Fox Populations.* Canadian Wildlife Service Report. Series no. 8, 1969.

Macpherson, Andrew H. "A Northward Range Extension of the Red Fox in the Eastern Canadian Arctic." *Journal of Mammalogy,* vol. 45, no. 1, 1964.

Manning, Th. "Remarks on the Reproduction, Sex Ratio, and Life Expectancy of the Varying Lemming." *Arctic,* vol. 7, no. 1, June 1954, Arctic Institute of North America.

Mansfield, A. W. "Distribution of the Harbor Seal, *Phoca vitulina linnaeus,* in Canadian Arctic Waters." *Journal of Mammalogy,* vol. 48, no. 2, 1967.

Mansfield, A. W. "The Grey Seal in Eastern Canadian Waters." *Canadian Audubon Magazine.* Canadian Audubon Society, 1966.

Mansfield, A. W. "The Mammals of Sable Island." The

Canadian Field-Naturalist, vol. 81, no. 1.

Mansfield, A. W. "Present Status of the Walrus Population at Southampton and Coats Islands." Unpublished manuscript.

Mansfield, A. W. *Seals of Arctic and Eastern Canada.* Fisheries Research Board of Canada, 1967.

Mansfield, A. W. "The Walrus in Canada's Arctic." *Canadian Geographical Journal*, vol. 72, no. 3, March 1966.

Marsden, Walter. *The Lemming Year.* Chatto and Windus, London, 1964.

Matthews, Leonard H. *The Whale.* Simon and Schuster, New York, 1968.

Maxwell, Gavin. *Seals of the World.* Constable, London, 1967.

Nansen, Fridtjof. *Farthest North*, 2 vols. Macmillan, London, 1897.

Nansen, Fridtjof. *In Northern Mists.* W. Heinemann, London, 1911.

Nansen, Fridtjof, and Collett, Robert. An account of the birds in *The Norwegian North Pole Expedition 1893-1896.* Longmans, Green & Co., London, 1900.

Peary, Robert E. *The North Pole.* Frederick A. Stokes, 1911.

Pedersen, Alwin. *Der Eisbär.* Ziemsen Verlag, Wittenberg, 1957.

Pedersen, Alwin. *Der Eisfuchs.* Ziemsen Verlag, Wittenberg.

Pedersen, Alwin. *Der Moschusochs.* Ziemsen Verlag, Wittenberg, 1958.

Pedersen, Alwin. *Polar Animals.* Harrap, London, 1958, 1962.

Pedersen, Alwin. *Das Walross.* Ziemsen. Wittenberg, 1962.

Perry, Richard. *The World of the Polar Bear.* University of Washington Press, Seattle, 1966.

Perry, Richard. *The World of the Walrus.* Taplinger Publishing Co., New York, 1967.

Peterson, Randolph L. *The Mammals of Eastern Canada.* Oxford University Press, Toronto 1966.

Porsild, A. E. *Illustrated Flora of the Canadian Arctic Archipelago.* National Museum of Canada, 1957.

Porsild, Morten. "On 'Savssats': A Crowding of Arctic Animals at Holes in the Sea Ice." *The Geographical Review*, vol. 6, no. 3, September 1918.

Rasmussen, Birger. On the Stock of Hood Seals in the Northern Atlantic. Fisheries Research Board of Canada, Translation Series no. 387, Montreal 1962.

Rutilevski, G. L. *A Narwhal in the Region of Drifting Station North Pole 5.* American Meteorological Society, Translation 182. Cambridge, Mass. 1958.

Salomonsen, Finn. *Grønlands Fugle:* The Birds of Greenland. Munksgaard, Copenhagen, 1950.

Scheffer, Victor B. *Seals, Sea Lions and Walruses.* Stanford University Press, 1958.

Scherman, Katherine. *Spring on an Arctic Island.* Little, Brown, Boston, 1956.

Sergeant, D. E. *The Biology and Hunting of Beluga or White Whales in the Canadian Arctic.* Fisheries Research Board of Canada, Arctic Biological Station, 1962.

Sergeant, D. E. "Exploitation and Conservation of Harp and Hood Seals." *The Polar Record*, vol. 12, no. 80.

Sergeant, D. E. "Migrations of Harp Seals, *Pagophilus groenlandicus (Erxleben)* in the Northwest Atlantic." *Journal Fisheries Research Board of Canada*, vol. 22, no. 2, 1965.

Sergeant, D. E. "Reproductive Rates of Harp Seals, *Pagophilus groenlandicus (Erxleben).*" *Journal Fisheries Research Board of Canada*, vol. 23, no. 5, 1966.

Sergeant, D. E., and Brodie, P. F. "Body Size in White Whales, *Delphinapterus leucas.*" *Journal Fisheries Research Board of Canada*, vol. 26, no. 10, 1969.

Sergeant, D. E., and Brodie, P. F. "Tagging White Wales in the Canadian Arctic." *Journal Fisheries Research Board of Canada*, vol. 26, no. 8, 1966.

Sergeant, D. E., and Fisher, H. D. *A Review of the Harp Seal Problem.* Fisheries Research Board of Canada, 1954.

Seton, Ernst T. *The Arctic Prairies.* International Universities Press, New York, 1911, 1948.

Slijper, E. *Whales.* Hutchinson, London, 1962.

Snyder, Lester L. *Arctic Birds of Canada.* University of Toronto Press, 1957.

Soper, J. D. "Discovery of the Breeding Grounds of the Blue Goose, 1929." *The Canadian Field-Naturalist*, vol. 44, no. 1, Jan. 1930.

Stefansson, Vilhjalmur. *The Friendly Arctic.* Macmillan, New York, 1921.

Stefansson, Vilhjalmur. *My Life with the Eskimos.* Macmillan, New York, 1913.

Sverdrup, Otto N. *New Land*, translated by Ethel Harriet Hearn, 2 vols. Longmans, Green, and Company, 1904.

Symington, Fraser. *Tuktu.* Canadian Wildlife Service, 1965.

Taylor, Jr., William E. *A Description of Sadlermiut Houses Excavated at Native Point, Southampton Island, N.W.T.* National Museum of Canada Bulletin 162, 1957.

Tener, J. S. *Muskoxen.* Canadian Wildlife Service Report Series no. 2, 1965.

Todd, W. E. *Birds of the Labrador Peninsula and Adjacent Areas.* University of Toronto Press, 1963.

Tuck, Leslie M. *The Murres.* Canadian Wildlife Service Report Series no. 1, 1961.

Tyrrell, James W. *Across the Sub-Arctics of Canada* (3rd edition). William Briggs, Toronto, 1908.

Uspenski, S. *The Bird Bazaars of Novaya Zemlya.* Russian game reports, vol. 4. Canadian Wildlife Service, Ottawa, 1958.

Wilkinson, Douglas. *Land of the Long Day.* Henry Holt & Co., New York, 1956.

The text material in this book
is set in 11 pt Melior Roman
and the chapter titles in Bullfinch Oldstyle.

The book was designed by Brant Cowie
and produced in Toronto:
lithography by Southam Murray
and binding by The Hunter Rose Company.

*(Overleaf) Arctic owl and
ptarmigan by Pauloosie Sivua
of Povungnituk.*